INDEPENDENT LEARNING PROJECT FOR ADVANCED CHEMISTRY

# ILPAC
## second edition

*INORGANIC*

## P-BLOCK ELEMENTS

REVISED BY ANN LAINCHBURY JOHN STEPHENS ALEC THOMPSON

**JOHN MURRAY**

# ■ ACKNOWLEDGEMENTS

We are grateful to CLEAPSS/ASE Laboratory Standards Committee for ensuring that the text meets with current safety recommendations.

Thanks are due to the following examination boards for permission to reproduce questions from past A-level papers:
Associated Examining Board: Exercises 26, p. 21 (1979); 51, p. 37 (1980); 70, p. 64 (1980); Teacher-marked exercise 3, p. 71 (1996 specimen); End-of-unit test 10, p. 77 (1990). Joint Matriculation Board: Teacher-marked exercises, p. 12 (1977); p. 34 (1979); p. 41 (1978); Part A test 3, p. 47 (1993); 9, p. 51 (1990); End-of-unit test 8, p. 76 (1991); 11, p. 77 (1992). Oxford and Cambridge Schools Examination Board: Teacher-marked exercise, p. 40 (1979); Exercise 79, p. 70 (1992); Part A test 6, p. 49 (1993); End-of-unit test 4, p. 75 (1992); 5, p. 75 (1990); 9, p. 77 (1992); 12, p. 78 (1992); 15, p. 78 (1994); Southern Examining Group: Exercises 22, p. 19 (1976); 39, p. 30 (1975); 52, p. 37 (1975). University of Cambridge Local Examinations Syndicate: Exercise 33, p. 24 (1974); Teacher-marked exercise 4, p. 71 (1994); Part A test 4, p. 48 (1991); End-of-unit test 1, p. 74 (1994); 2, p. 74 (1992); 3, p. 74 (1994); 7, p. 76 (1992). University of London Examinations and Assessment Council: Exercises 4a, p. 10 (1979); 4b and 4c, p. 10 (1978); 10, p. 13 (N 1980); 23, p. 20 (N 1976); 37, p. 25 (N 1977); 65, p. 55 (N 1978); 72, p. 65 (N 1974); 73, p. 66 (N 1976); 74, p. 67 (1980); 77, p. 69 (1980); Part A test 1, p. 47 (1994); 2, p. 47 (1989); 5, p. 49 (1992); 7, p. 50 (1991); 10, p. 51 (N 1991); Teacher-marked exercises, p. 16 (1990); p. 42 (1988); p. 70 (1990); End-of-unit test 13, p. 78 (1992); End-of-unit practical test, p. 72 (1977); University of Oxford Delegacy and Local Examinations: Exercise 29, p. 22 (1981); Teacher-marked exercise 2, p. 71 (1993); Part A test 8, p. 50 (1993); End-of-unit test 14, p. 78 (1990).
(The examination boards accept no responsibility whatsoever for the accuracy or method of working in the answers given.)
Photographs reproduced with kind permission of Andrew Lambert (p. 14); Mike McNamee/ Science Photo Library (p. 36); Martin Bond/Science Photo Library (p. 58); ICI plc (p. 64). All other photographs by Last Resort Picture Library. The assistance provided by the staff and students of Djanogly City Technology College, Nottingham for the photographs of experiments is gratefully acknowledged.
The publishers have made every effort to trace copyright holders, but if they have inadvertently overlooked any they will be pleased to make the necessary arrangements at the earliest opportunity.

Original material produced by the Independent Learning Project for Advanced Chemistry sponsored by the Inner London Education Authority

First edition published in 1983
by John Murray (Publishers) Ltd
50 Albemarle Street
London W1X 4BD

Second edition 1996

Design and layouts by John Townson/Creation.
Illustrations by Barking Dog Art.

Typeset in 10/12 pt Times and Helvetica by Wearset, Boldon, Tyne and Wear.

Printed in Great Britain by St Edmundsbury Press Ltd, Bury St Edmunds.

*British Library Cataloguing in Publication Data*
A catalogue record for this book is available from the British Library.

ISBN 0–7195–5342–3

# CONTENTS

# ■ Symbols used in ILPAC

 Computer program

 A-level question

 Discussion

 A-level part question

 Experiment

 A-level question; Special Paper

 Model-making

A-level supplementary question

 Reading

 Revealing Exercise

 Video programme

# ■ International hazard symbols

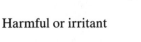 Corrosive

 Oxidising

Explosive

 Radioactive

 Harmful or irritant

 Toxic

 Highly flammable

# P-BLOCK ELEMENTS

# INTRODUCTION

In this book we deal with Groups III, V and VI, as shown in the outline Periodic Table below.

**Figure 1**

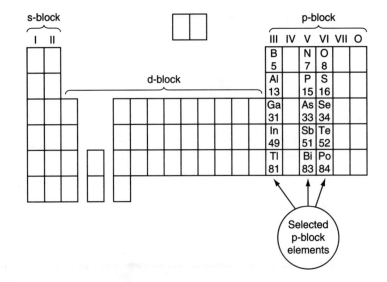

The elements in these groups show the usual gradation from non-metallic to metallic properties with increasing atomic number.

The chemistry of the first two elements in each group will be treated in detail in Part A. We include a brief account of the chemistry of the remaining elements in each group only in so far as it illustrates group trends.

In Part B we deal with the industrial extraction of some elements and the manufacture of some compounds from these groups.

The book as a whole covers a great deal of chemistry; it is unlikely that you will need to study all of it. Consult your teacher if you are not clear about what is required for your syllabus.

There are five experiments in this book, three of them in Part A, one in Part B and one as an end-of-unit practical test.

## ■ Pre-knowledge

Before you start work on this book you should be able to:
■ Explain the terms:
    **a** amphoteric,
    **b** lattice enthalpy, or lattice energy,
    **c** electronegativity,
    **d** oxidation number,
    **e** hydrogen bonding,
    **f** polarisation of both bonds and ions.
■ Describe the action of heat and of an alkali on ammonium chloride.
■ Describe a test for sulphate ions in solution.
■ Outline the preparation of hydrogen peroxide.
■ Deduce the shapes of molecules by means of the electron-pair repulsion theory.
■ Explain why solutions of iron(III) salts are acidic.
■ Use standard electrode potentials to predict the outcome of chemical reactions and to write conventional cell diagrams.
■ Use Le Chatelier's principle to predict the effect on an equilibrium system of changing the conditions.
■ Perform calculations using the expression for the equilibrium constant, $K_p$, for a chemical reaction.

## ■ Pre-test

To find out whether you are ready to start Part A try the following test, which is based on the pre-knowledge items. You should not spend more than 60 minutes on the test. Hand your answers to your teacher for marking.

1. Explain, by reference to the mutual repulsion of electron pairs, the shapes of the molecules of ammonia, $NH_3$, and boron trifluoride, $BF_3$. How do these substances react together? (6)

2. Zinc oxide, $ZnO$, is amphoteric. Explain this statement, illustrating your answer with two ionic equations. (3)

3. When a white solid, X, is heated with dilute sodium hydroxide solution, a gas, Y, is evolved which turns damp red litmus blue. An aqueous solution of X gives a white precipitate with silver nitrate solution.
   **a** Identify X and Y and give equations for the two reactions mentioned. (4)
   **b** Describe, with an equation, what happens when X is heated alone. (3)

4. Outline the preparation of aqueous hydrogen peroxide, starting from barium metal. (3)

5. The lattice energy of magnesium chloride is $-2489 \text{ kJ mol}^{-1}$. Write the relevant thermochemical equation. (2)

6. What two structural features must a substance possess in order to form hydrogen bonds between its molecules? (3)

7. Hydrogen fluoride is usually regarded as covalent, but it shows a significant degree of ionic character. Explain this statement briefly, using the terms 'electronegative' and 'polarisation' in your answer. (3)

8. Calculate the oxidation number of nitrogen in each of the following:
   **a** $NH_3$
   **b** $HNO_3$
   **c** $NO_2^-$ (3)

9. What is the formula of the most numerous of the complex ions in a solution of iron(III) nitrate in water? Explain briefly why this solution is acidic. (4)

10. Describe a simple test for sulphate ions in solution. Write an equation. (3)

11. Standard electrode potentials for two half-cells are as follows:

$$Cd^{2+}(aq) + 2e^- \rightleftharpoons Cd(s) \qquad E^{\ominus} = -0.40 \text{ V}$$
$$Ag^+(aq) + e^- \rightleftharpoons Ag(s) \qquad E^{\ominus} = +0.80 \text{ V}$$

   **a** Which is more readily oxidised, cadmium or silver? (1)
   **b** Deduce an equation for the reaction which occurs between one of the metals and the aqueous ions of the other metal. (1)
   **c** Write a conventional cell diagram for a cell in which the reaction in (b) occurs. What is its e.m.f.? (State the sign.) (2)

12. The formation of ammonia is represented by the thermochemical equation:

$$N_2(g) + 3H_2(g) \rightleftharpoons 2NH_3(g) \qquad \Delta H^\ominus = -94.6 \text{ kJ mol}^{-1}$$

    **a** Use Le Chatelier's principle to deduce the effect on this equilibrium of increasing
       i) pressure, and
       ii) temperature. (4)
    **b** Write an expression for the equilibrium constant, $K_p$. (1)
    **c** A mixture of hydrogen and nitrogen in the molar ratio $3:1$ was allowed to reach equilibrium at 100 atm and 400 °C. The equilibrium mixture contained 25% ammonia. Calculate the value of $K_p$ under these conditions. (4)

                              (Total: 50 marks)

# THE ELEMENTS AND SOME COMPOUNDS

# CHAPTER 1

# GROUP III: BORON AND ALUMINIUM

In ILPAC 4, The s-Block Elements and The Halogens, you learned that the top element in each group is atypical. Similarly, in Group III, you might expect boron to differ significantly from aluminium.

In the first section we consider the properties of the elements in Group III, with particular reference to comparisons between boron and aluminium.

## ■ 1.1 The nature of the elements

Boron and aluminium are too reactive to be found free in nature. Boron occurs principally in borates, the most important being sodium tetraborate (borax), $Na_2B_4O_7 \cdot 10H_2O$, and as borosilicates. Aluminium occurs in a variety of aluminosilicates and its hydrated oxide. It is the most abundant metal in the Earth's crust, as shown in Fig. 2.

**Figure 2**
Composition of the Earth's crust.

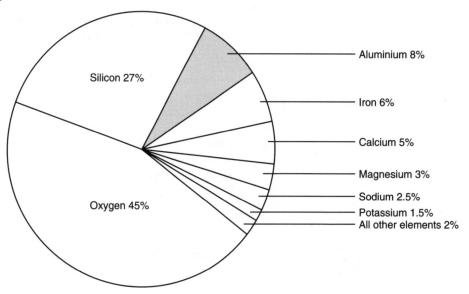

- Aluminium 8%
- Silicon 27%
- Iron 6%
- Calcium 5%
- Magnesium 3%
- Sodium 2.5%
- Potassium 1.5%
- All other elements 2%
- Oxygen 45%

When you have finished this section you should be able to:
- compare **physical properties of boron and aluminium**, such as appearance, conductivity and ionic radii;
- list the likely **oxidation states** of the elements in Group III;
- state how aluminium reacts with **acids and alkalis**;
- explain how **aluminium sulphate** can be prepared from aluminium.

To obtain a simple overall picture of Group III, we suggest that you read the introduction to these elements in your textbook(s). Pay particular attention to their appearance, their electrical conductivity and the ores from which boron and aluminium are extracted. This will help you to do the following exercises.

**EXERCISE 1**
*Answers on page 81*

**a** Name the ore from which aluminium is extracted. What compound is the main constituent of this ore?
**b** Describe the appearance and electrical conductivity of boron and aluminium.
**c** What does your answer to (b) suggest about the metallic character of these two elements?
**d** What would you expect to be the appearance and electrical conductivity of thallium?

We discuss the extraction of aluminium from its ore in detail in Part B. The next exercise is related to the atomic properties of Group III elements.

**EXERCISE 2**
*Answers on page 81*

**a** Write the electronic configurations of boron and aluminium using the s, p, d notation.
**b** What oxidation states would you expect boron and aluminium to exhibit?
**c** Would you expect to encounter the 'inert pair effect' in this group? Explain your answer.
  List the possible oxidation states of thallium.
**d** i) Record the ionic radii of boron and aluminium from your data book. How do they compare with other ionic radii?
  ii) Would you expect to find many compounds containing $B^{3+}$ or $Al^{3+}$ ions? Explain, with reference to your answer to i).

In the next exercise you make some predictions about the reactions of aluminium by referring to standard electrode potentials. These reactions are important in relation to the extensive use of aluminium in manufacturing.

**EXERCISE 3**
*Answers on page 81*

By reference to the standard electrode potentials, $E^{\ominus}$, listed below, predict whether aluminium will react with:
**a** dilute hydrochloric acid,
**b** dilute sodium hydroxide solution.

| **Electrode reaction** | $E^{\ominus}$/**V** |
|---|---|
| 1. $Al^{3+}(aq) + 3e^- \rightleftharpoons Al(s)$ | −1.66 |
| 2. $Al(OH)_4^-(aq) + 3e^- \rightleftharpoons Al(s) + 4OH^-(aq)$ | −2.35 |
| 3. $2H^+(aq) + 2e^- \rightleftharpoons H_2(g)$ | 0.00 |
| 4. $2H_2O(l) + 2e^- \rightleftharpoons H_2(g) + 2OH^-(aq)$ | −0.83 |

You can test your predictions in the next experiment.

## EXPERIMENT 1    Reactions of aluminium

**Aim**    The purpose of this experiment is to show how aluminium foil reacts with dilute acids and alkalis and, when reactions do occur, to carry out tests on the resulting solutions.

**Introduction**    In ILPAC 4, The Periodic Table, you learned that aluminium oxide is amphoteric. Other metals with amphoteric oxides such as beryllium and zinc tend to react with acids and alkalis, so you might also expect aluminium to react in a similar way.

In addition to observing the reactions of aluminium with dilute acids and alkalis, you also attempt to interpret some reactions of the resulting solutions.

**Requirements**    ■ safety spectacles
■ 6 pieces of aluminium foil, 3 cm × 3 cm
■ 10 test-tubes
■ test-tube rack
■ test-tube holder
■ sodium hydroxide solution, 2 M NaOH
■ hydrochloric acid, dilute, 2 M HCl
■ Bunsen burner and bench mat
■ beaker, 100 cm$^3$
■ copper(II) chloride solution, 0.1 M CuCl$_2$
■ mercury(II) chloride solution, 0.1 M HgCl$_2$

- wash-bottle of distilled water
- filter funnel
- filter paper
- sulphuric acid, dilute, 1 M $H_2SO_4$
- sodium carbonate solution, 1 M $Na_2CO_3$
- red and blue litmus papers

## HAZARD WARNING

Sodium hydroxide solution is corrosive, even in dilute solution.

Mercury(II) chloride is highly toxic by ingestion and by skin absorption.
Therefore you **must**:
- **wear safety spectacles and protective gloves;**
- **work with great care.**

(Your teacher may choose to demonstrate the use of mercury(II) chloride in this experiment.)

**Procedure – Part A**

### Reactions of aluminium foil

1. Tear one of the sheets of aluminium foil into smaller pieces and place them in a test-tube containing 3–4 $cm^3$ of dilute sodium hydroxide solution. Identify any gas given off and record your observations in a copy of Results Table 1. Keep the resulting solution for further tests.
2. Repeat step 1 using dilute hydrochloric acid instead of sodium hydroxide solution. If no reaction occurs, heat the mixture gently.
3. Place 3 pieces of aluminium in a small beaker and just cover them with copper(II) chloride solution. On a fourth piece of aluminium, place 3 separate drops of mercury(II) chloride solution.
4. After about 2 minutes, pour away the copper(II) chloride solution and rinse the foil with distilled water. Leave one piece exposed to the air. Put the others in 2 test-tubes and add sodium hydroxide and dilute hydrochloric acid respectively. Compare what happens with the results of steps 1 and 2.
5. Rinse away the drops of mercury(II) chloride solution with distilled water and leave the foil exposed to the air. Examine it after a few minutes and compare it with the foil treated with copper(II) chloride.

**– Part B**

### Reactions of the resulting solutions

6. Filter the resulting solutions from steps 1 and 2 (if necessary) and test the filtrates as follows:
7. To separate 1 $cm^3$ portions of the filtrates from step 6 add, as appropriate, the following reagents, a little at a time, until they are present in excess:
   **a** dilute sulphuric acid (to the solution in NaOH only),
   **b** dilute sodium hydroxide solution (to the solution in HCl only),
   **c** sodium carbonate solution (to the solution in HCl only).
8. Record your observations in a copy of Results Table 2. If no reaction occurs write **none** in the appropriate box.

**Results Table 1**
Reaction of aluminium foil

| Reagent | Observations | Identity of any gas given off |
|---|---|---|
| Sodium hydroxide solution | | |
| Sodium hydroxide solution after immersion in $CuCl_2(aq)$ | | |
| Dilute hydrochloric acid | | |
| Dilute hydrochloric acid after immersion in $CuCl_2(aq)$ | | |
| Air after immersion in $CuCl_2(aq)$ | | |
| Air after immersion in $HgCl_2(aq)$ | | |

**Results Table 2**
Reactions of the solutions

| Reagent | Observations using solutions of | |
|---|---|---|
| | Al in NaOH | Al in HCl |
| Dilute sulphuric acid | | |
| Sodium hydroxide solution | | |
| Sodium carbonate solution | | |

*Specimen results on page 82*

You may need to refer to your textbook(s) in order to answer the following questions.

Aluminium has a wide range of uses.

**Questions**

*Answers on page 82*

1. Untreated aluminium has a wide variety of uses which depend, in part, on its resistance to corrosion in normal conditions. However, standard electrode potentials suggest that aluminium is more reactive than iron, which corrodes badly. Explain briefly. (You study the uses of aluminium further in Part B.)
2. How does treatment with a solution of copper(II) chloride or mercury(II) chloride reveal the true reactivity of aluminium?
3. Suggest a reason for the fact that mercury(II) chloride is more effective than copper(II) chloride in promoting a reaction between aluminium and air.
4. Why should you not use washing soda or some special oven-cleaners on aluminium kitchenware?
5. Use the data in Exercise 3 to write equations for the reactions of aluminium with
   **a** dilute acids      **b** alkalis.
6. Explain, as far as possible, the observations you have made in Results Table 2 and name the precipitates formed. (Hint: aluminium carbonate is not known.)

Now attempt the next exercise, which is related to the experiment you have just done and also helps you to revise calculations based on the mole.

**EXERCISE 4**

*Answers on page 83*

**a** Suggest how the compound $Al_2(SO_4)_3 \cdot xH_2O$ could be prepared in crystalline form from aluminium.
**b** A hydrated aluminium sulphate, $Al_2(SO_4)_3 \cdot xH_2O$, contains 8.1% of aluminium by mass.
   i)   calculate the value of $x$;
   ii)  comment on the colour of an aqueous solution of the salt.
**c** The standard electrode potential for the couple $Al^{3+}(aq)/Al(s)$ is $-1.66$ V. Why does aluminium not dissolve readily in dilute acids?

In ILPAC 3, Bonding and Structure, you learned that there are compounds in which boron and aluminium atoms have fewer than eight electrons in their outer shells. We now consider some of these compounds, which are often described as being 'electron deficient'.

## ■ 1.2 The halides of boron and aluminium

The halides of boron and aluminium may be prepared by direct combination. You have already prepared aluminium chloride by this method in ILPAC 4, The Periodic Table. In this section you **build upon the knowledge of these compounds you have gained** from ILPAC 3, Bonding and Structure.

**OBJECTIVE**   When you have finished this section you should be able to:
■ describe the bonding and structure of the **fluorides and chlorides of boron and aluminium**.

In the next exercise you consider the bonding in some of the halides.

**EXERCISE 5**
*Answers on page 84*

a Look up and record the boiling points of the fluorides and chlorides of boron and aluminium.
b Which of the above halides has the greatest ionic character? Explain your answer.
c Place the aluminium halides in order of decreasing ionic character and explain your answer in terms of polarisability of the anion.

In ILPAC 4, Bonding and Structure, you learned how to predict the shapes of molecules and recognise ionic and covalent substances. In the next exercise you apply this knowledge to the chlorides of boron and aluminium.

**EXERCISE 6**
*Answers on page 84*

a Draw a diagram of the $BCl_3$ molecule, indicating the shape and bond angles.
b Would you expect the chlorides of boron and aluminium to dissolve in organic solvents? Explain your answer.
c With the aid of diagrams and equations, explain the following facts.
  i)  The relative molecular mass ($M_r$) of aluminium chloride vapour below 400 °C is 267.0 whereas at 800 °C it is 133.5.
  ii) At 600 °C, $M_r$ for aluminium chloride vapour is between 267.0 and 133.5.
d Explain why $M_r$ for boron chloride vapour does not appreciably exceed 117. (Hint: consider the small size of the boron atom.)

You learned in ILPAC 4, The Periodic Table, that the chlorides of boron and aluminium are hydrolysed by water. The same is true of the bromides and iodides, but not the fluorides. The next exercise concerns the hydrolysis of the chlorides of boron and aluminium. You may need to refer to your textbooks for some of the details.

**EXERCISE 7**
*Answers on page 85*

Study the following equations for hydrolysis reactions:

$$BCl_3(l) + 3H_2O(l) \rightarrow B(OH)_3(aq) + 3HCl(aq)$$
$$AlCl_3(s) + 6H_2O(l) \rightarrow [Al(H_2O)_6]^{3+}(aq) + 3Cl^-(aq)$$

a Copy the above equations and name the products formed in each reaction.
b Are the resulting solutions acidic, basic or neutral?
c Suggest a reason for the fact that the chloride of the top element in Group IV, $CCl_4$, does not hydrolyse whereas $BCl_3$ does.
d Sketch the shape of the complex ion $[Al(H_2O)_6]^{3+}$.
e Why does boron **not** form six co-ordinate complexes like the one you have just described?
f What alternative molecular formula is often used for $B(OH)_3$?

The last exercise suggests that $B(OH)_3(aq)$ and $[Al(H_2O)_6]^{3+}(aq)$ behave as acids. We explore this in the next section.

# ■ 1.3  Acidity of boron and aluminium compounds

When you have finished this section you should be able to:
■ describe the **hydrolysis** of the chlorides of boron and aluminium;
■ explain the acidic nature of solutions of **boric(III) acid**, $B(OH)_3$, and **aluminium salts**.

In ILPAC 11, Transition Elements, you learned that aqua complexes are often acidic, especially when the metal ions are small and highly charged. If necessary, refer back to the section on acidity of complex ions in ILPAC 11, to help you with the next exercise.

**EXERCISE 8**
*Answers on page 85*

In aqueous solution, $B(OH)_3$ forms hydronium (oxonium) ions as represented by the following equation:

$$B(OH)_3(aq) + 2H_2O(l) \rightarrow [B(OH)_4]^-(aq) + H_3O^+(aq)$$

**a** Write equations to show the formation of hydronium ions by $[Al(H_2O)_6]^{3+}$.
**b** Look up and record values of $K_a$ for the species $B(OH)_3$ and $[Al(H_2O)_6]^{3+}$. Which is the stronger acid?
**c** Explain why $B(OH)_3$ reacts with a water molecule to form $[B(OH)_4]^-(aq)$.
**d** The dissociation of the hydrated aluminium(III) ion suggests that the O—H bond in the coordinated water molecule is weakened so that an adjacent uncoordinated water molecule can more readily accept a proton from it. Why do you think this might happen?
**e** Why is the transfer of a proton from $[Al(H_2O)_5(OH)]^{2+}$ less likely than from $[Al(H_2O)_6]^{3+}$ in aqueous solution?

The next exercise investigates the conditions under which three or more protons can be transferred from $[Al(H_2O)_6]^{3+}$. It is related to Experiment 1.

**EXERCISE 9**
*Answers on page 86*

In the hydrolysis of the hexaaquaaluminium(III) ion $[Al(H_2O)_6]^{3+}$, the water behaves as a weak base. The product of hydrolysis depends on the strength of the base. With this in mind, interpret the following observations:
**a** When sodium carbonate solution is added to $[Al(H_2O)_6]^{3+}$ a white precipitate forms.
**b** When sodium hydroxide solution is added to $[Al(H_2O)_6]^{3+}$ a white precipitate forms initially but this re-dissolves in excess sodium hydroxide solution.

In the following teacher-marked exercise you compare the acidity of solutions of aluminium ions and other metal ions. It would be instructive to include sodium ions (radius 0.095 nm) in your discussion.

**EXERCISE**
*Teacher-marked*

Discuss the relative acidity of solutions of $AlCl_3$, $FeCl_2$ and $FeCl_3$, given the ionic radii $Al^{3+}$, 0.050 nm; $Fe^{2+}$, 0.076 nm; $Fe^{3+}$, 0.065 nm.

The hexaaquaaluminium(III) ion yields a species having the general formula $[Al(H_2O)_x(OH)_y]^z$ when treated with a base. State and justify a simple algebraic expression connecting:
i) $x$ and $y$,
ii) $y$ and $z$.

Show why the particular species obtained is related to the strength of the base added, using $H_2O$, $CO_3^{2-}$, and $OH^-$ as examples of bases.

How are the foregoing principles related to the observations made when 2 M sodium hydroxide is added to solutions of
**a** $AlCl_3$,
**b** $FeCl_2$,
**c** $FeCl_3$?

Having discussed the halides we now consider some Group III oxides.

## ■ 1.4 The oxides of boron and aluminium

In ILPAC 4, The Periodic Table, you learned that boron oxide is acidic and has a giant covalent structure whereas aluminium oxide is amphoteric and has a giant ionic structure. These properties are consistent with the change from non-metal in the case of boron to 'weak' metal in the case of aluminium. The word 'weak' is used because aluminium does not show all the typical chemical characteristics of metals.

**OBJECTIVES**

When you have finished this section you should be able to:
■ write a chemical equation for the reaction between **boron oxide and water**;
■ list some commercial **uses of aluminium oxide**;
■ compare the **stabilities of boron oxide and aluminium oxide**.

The next two exercises deal with boron oxide. The first uses the following data about boron and its compounds. On the graph the number of electrons removed is plotted against the logarithm of the appropriate ionisation energy and, in Table 1, standard enthalpy changes are listed for various processes.

**Figure 3**

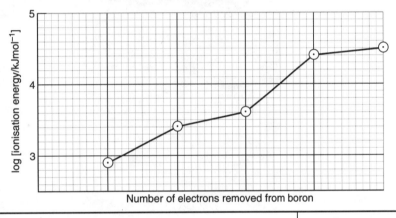

**Table 1**

| Process | $\Delta H^{\ominus}$(298 K)/kJ mol$^{-1}$ |
|---|---|
| First electron affinity of oxygen | −140 |
| Second electron affinity of oxygen | +790 |
| Standard enthalpy of atomisation of oxygen | +250 |
| Standard enthalpy of atomisation of boron | +590 |
| Standard enthalpy of formation of the oxide of boron | −1270 |
| First ionisation energy of boron | +800 |
| Second ionisation energy of boron | +2400 |
| Third ionisation energy of boron | +3700 |
| Fourth ionisation energy of boron | +25000 |
| Fifth ionisation energy of boron | +32800 |

**EXERCISE 10**

*Answers on page 86*

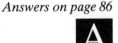

a On (a copy of) the graph, label each point with the electronic configuration (s, p, d or f) of the corresponding electron. Deduce the most likely formula for the oxide of boron.

b Draw a labelled Born–Haber cycle for the formation of the oxide of boron (assumed to be ionic). Calculate a lattice energy for the oxide of boron.

c One mole of the oxide of boron reacts with three moles of water and the product dissolves in water to give a weakly acidic solution. Write an equation, with state symbols, for the equilibrium occurring in aqueous solution.

d Would you expect your value for the lattice energy of the oxide of boron to be in good or poor agreement with a theoretically derived value? Give your reason(s).

**EXERCISE 11**

*Answers on page 87*

How do the oxides of boron and aluminium react with:
**a** sodium hydroxide solution,
**b** dilute hydrochloric acid?
Write ionic equations where appropriate. (For the products, use formulae you have already met in previous exercises.)

In the next exercise you compare the stabilities of aluminium oxide and boron oxide. You need not use $\Delta G^{\ominus}$ values if you have not studied them before.

**EXERCISE 12**

*Answers on page 88*

The thermit process shown here is used to mend/weld steel rails.

**a** Write down the values of $\Delta H_f^{\ominus}$ and/or $\Delta G_f^{\ominus}$ for $B_2O_3$ and $Al_2O_3$, using your data book.
**b** Which of the oxides is the more stable? Explain your answer.

In ILPAC 2, Chemical Energetics, you learned that aluminium will reduce iron(III) oxide to iron and form aluminium oxide (thermit process). This process makes use of the high negative enthalpy of formation of aluminium oxide.

 Read about the uses of aluminium oxide in your textbook(s) so that you can do the following exercise. We suggest that you leave until Part B a study of the process known as anodising, which thickens the protective layer of oxide on the surface of aluminium.

**EXERCISE 13**

*Answer on page 88*

List three important uses of aluminium oxide.

## ■ 1.5 Aluminium sulphate and potassium aluminium sulphate

**OBJECTIVES**

When you have finished this section you should be able to:
■ state the effect of **heat on aluminium sulphate**;
■ describe the **preparation of potassium aluminium sulphate** (potash alum);
■ state the **general formula of alums**;
■ describe how potassium aluminium sulphate reacts with dilute **sodium hydroxide**, **barium chloride** and **sodium carbonate** solutions;
■ list some **uses of aluminium sulphate**.

Crystal of potassium aluminium sulphate.

Read about the preparation of potassium aluminium sulphate, its properties, uses and formula. You should also read about the properties and uses of aluminium sulphate and compare it with potassium aluminium sulphate. This will help you with the two exercises that follow.

**EXERCISE 14**
*Answers on page 88*

a Write an equation to show the thermal decomposition of aluminium sulphate.
b Suggest a reason, in terms of the polarising effect of $Al^{3+}$, for the relative ease of decomposition of aluminium sulphate compared with other sulphates.
c State three uses of aluminium sulphate.

**EXERCISE 15**
*Answers on page 88*

a Briefly describe how crystals of potassium aluminium sulphate can be made from potassium sulphate and aluminium sulphate.
b State the general formula of the alums and give the formula of potassium aluminium sulphate.
c Would you expect a solution of potassium aluminium sulphate to be neutral, acidic or basic? Explain.

In the next exercise you compare the reactions of aluminium sulphate and potassium aluminium sulphate. You may find it useful to consult a textbook of inorganic qualitative analysis.

**EXERCISE 16**

*Answers on page 89*

**Table 2**

**a** Complete a copy of Table 2.

| Test | Observations | Equation(s) |
|---|---|---|
| 1. Flame test<br> i) $Al_2(SO_4)_3$<br> ii) $KAl(SO_4)_2$ |  |  |
| 2. Dropwise addition<br> of excess NaOH(aq)<br> i) $Al_2(SO_4)_3(aq)$<br> ii) $KAl(SO_4)_2(aq)$ |  |  |
| 3. Addition of $Na_2CO_3(aq)$<br> i) $Al_2(SO_4)_3(aq)$<br> ii) $KAl(SO_4)_2(aq)$ |  |  |
| 4. Addition of $BaCl_2(aq)$<br> i) $Al_2(SO_4)_3(aq)$<br> ii) $KAl(SO_4)_2(aq)$ |  |  |

**b** What is the difference between a double salt and a complex salt? In which category are the alums?

In order to consolidate your knowledge of Group III elements, attempt one or both of the following teacher-marked exercises. Don't forget to **plan** your answer before you start writing.

**EXERCISE**

*Teacher-marked*

**a** Give an account of the similarities and differences between boron and aluminium, illustrating your answer with references to:
 i)   the physical properties of the elements,
 ii)  the stability and acid–base nature of the oxides,
 iii) the bonding and structure of the halides,
 iv)  the hydrolysis of the chlorides.
**b** Give **one** illustration in each case of the diagonal relationship of boron and aluminium with elements in neighbouring groups.
**c** Suggest **two** important ways in which you would expect thallium compounds to differ from boron compounds.

**EXERCISE**

*Teacher-marked*

**a** Give an account of the chemistry of boron and aluminium, considering particularly their oxides and chlorides. Structure your answer so as to contrast the essentially non-metallic nature of boron with the essentially metallic nature of aluminium. Explain why the elements behave so differently.
**b** 'Alums' are salts in which a 1+ cation and a 3+ cation crystallise in a lattice with sulphate ions. Their general formula is $X^+Y^{3+}(SO_4^{2-})_2 \cdot 12\ H_2O$, where $X^+$ and $Y^{3+}$ are the two cations. Y is often aluminium but need not be.
 i)   Explain why, in water, alums give positive tests for $X^+$, $Y^{3+}$ and $SO_4^{2-}$ ions. How would you show this by simple tests on ammonium aluminium sulphate?
 ii)  Chrome alum is potassium chromium(III) sulphate. One method of preparing it is to reduce a solution of potassium dichromate(VI) in dilute sulphuric acid with sulphur dioxide. Give half-equations for the two redox processes involved and hence give a balanced ionic equation for the reaction. Use this equation to show how the preparation leads to the required alum needing no separation from other products.

Apart from industrial aspects, which you consider in Part B, you have now completed a study of Group III elements. Ask your teacher how much of the next chapter, which concerns Group V, is relevant to your syllabus.

# GROUP V: NITROGEN AND PHOSPHORUS

By now, you have come to expect considerable differences between the first and second members of a group in the Periodic Table. In Group V, the contrast appears to be very great since nitrogen is a colourless, unreactive gas while phosphorus, especially in its white form, is a very reactive solid. However, you will find that the usual group trends and similarities are also evident, particularly in the compounds.

Nitrogen constitutes four-fifths of the earth's atmosphere and it also occurs in chemical combination, mainly as nitrates. It is an essential element to living matter, being an important constituent of proteins and nucleic acids. Although white phosphorus is toxic, phosphorus compounds are necessary for life processes; the remaining elements in the group, especially arsenic, are toxic in elemental or combined forms towards most living organisms.

## ■ 2.1 The nature of the elements

**OBJECTIVES**

When you have finished this section you should be able to:
- ■ describe the **structures of the red and white allotropes** (polymorphs) of **phosphorus**;
- ■ explain why **nitrogen** is a fairly unreactive gas while **phosphorus** is a reactive solid;
- ■ describe the **reactions of nitrogen and phosphorus** with:
  - **a** hydrogen,
  - **b** oxygen,
  - **c** chlorine,
  - **d** selected metals,
  - **e** alkali, and
  - **f** concentrated nitric acid.

Read the introduction to Group V in your textbooks, paying particular attention to the allotropes of phosphorus and their structure, and to the chemical reactions of nitrogen and phosphorus. Use the information you find to help you with the following exercises.

**EXERCISE 17**
*Answers on page 89*

**a** Write the electronic configurations of nitrogen and phosphorus atoms using the s, p, d notation. How many outer-shell electrons are there in all Group V elements?
**b** Draw a dot-and-cross diagram to show the bonding in the $N_2$ molecule.
**c** Name three allotropes of phosphorus and describe their appearance.
**d** Draw diagrams showing how the phosphorus atoms are linked in red and white phosphorus. Which has the higher boiling point?
**e** How may red phosphorus be converted into white phosphorus?
**f** Bearing in mind the usual group trends, what would you expect to be the structure of bismuth?

In the next exercise you are asked to make some predictions about ion formation in Group V from your knowledge of other groups in the Periodic Table.

**EXERCISE 18**
*Answers on page 90*

**a** Generally speaking, the compounds of all Group V elements have considerable covalent character. Which of the elements (if any) are most likely to form the following ions? (X denotes any Group V element.)
  i) $X^{5-}$,     ii) $X^{3-}$,     iii) $X^{3+}$.
**b** Name one compound that contains the $X^{3-}$ ion and one that contains the $X^{3+}$ ion.

The next exercise concerns a practical situation. You should not attempt the experiment because it is potentially dangerous.

**EXERCISE 19**

*Answers on page 90*

A few pieces of white phosphorus heated with a solution of sodium hydroxide produce a colourless gas called phosphine, PH$_3$. The apparatus shown in Fig. 4 below can be used to carry out this reaction but, before heating, it is necessary to pass a slow stream of nitrogen through the apparatus for a few minutes.

**Figure 4**

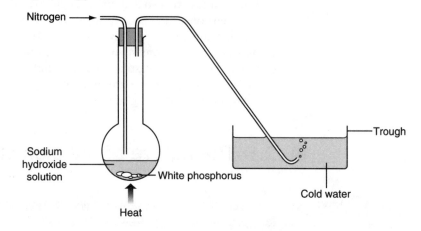

Phosphine has an objectionable odour, is highly poisonous and is spontaneously flammable when in contact with air. It is only very slightly soluble in water and has no effect on moist pH paper.

**a** Explain why it is necessary to pass nitrogen through the apparatus before heating the flask.

**b** What would you expect to happen when the phosphine gas bubbles reach the surface of the water in the trough?

**c** Describe how you would collect a sample of the phosphine gas prepared in this way.

**EXERCISE 20**

*Answers on page 90*

**a** Complete a larger copy of Table 3 which summarises the reactions of nitrogen and phosphorus.

**Table 3**

| Reagent | Nitrogen | White phosphorus | Red phosphorus |
|---------|----------|------------------|----------------|
| Hydrogen |  | No reaction | No reaction |
| Oxygen |  |  | Ignites at 260 °C → P$_4$O$_6$ or P$_4$O$_{10}$ |
| Chlorine | No reaction |  |  |
| Magnesium |  |  |  |
| Alkali | No reaction |  | No reaction |
| Conc. HNO$_3$ | No reaction |  |  |

**b** Write equations for the reactions between white phosphorus and:
  i) sodium hydroxide solution,
  ii) magnesium.
**c** State one chemical and one physical difference (in addition to colour) between red and white phosphorus.
**d** What is the reason for the differences in (c)?

In the last exercise you saw that nitrogen is fairly unreactive while phosphorus, especially the white allotrope, is very reactive. (We deal with the exception to this general rule, namely the reactivity towards hydrogen, later in this book.)
  In the next exercise you investigate the reasons why phosphorus is generally more reactive than nitrogen.

**EXERCISE 21**

*Answers on page 91*

Figure 5 shows some bond energy terms for Group V elements. Use the data to answer the questions that follow.

**Figure 5**

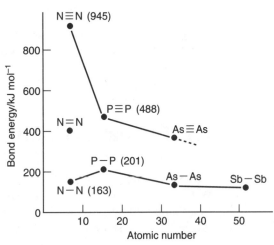

**a** State the type and number of bonds in a $P_4$ molecule. (Refer to Exercise 17, if necessary.)
**b** Calculate $\Delta H^\ominus$ for the following processes:
  i)   $4P(g) \rightarrow P_4(g)$
  ii)  $4N(g) \rightarrow N_4(g)$
  iii) $4P(g) \rightarrow 2P\equiv P(g)$
  iv)  $4N(g) \rightarrow 2N\equiv N(g)$
**c** In view of your answer to (b), suggest why nitrogen forms diatomic molecules whereas P, As and Sb in the vapour state all exist as $X_4$ molecules.
**d** Why do you think the $N\equiv N$ bond is so much stronger than the $P\equiv P$ bond? (Hint: Consider how multiple bonds are formed by the overlap of atomic orbitals.)
**e** Suggest a reason why nitrogen is less reactive than phosphorus.

With the above exercise in mind attempt the next exercise, which is part of an A-level question.

**EXERCISE 22**

*Answer on page 92*

Consider the structures of molecular nitrogen and phosphorus and attempt to explain the relative affinities of these two elements towards:
**a** oxygen,
**b** chlorine.

The next exercise is an A-level question that deals with the reaction of white phosphorus with copper(II) sulphate solution. You are not expected to know this reaction – you work out the details in the exercise.

**EXERCISE 23**

*Answers on page 92*

White phosphorus reacts with dilute aqueous solutions of copper(II) sulphate to deposit metallic copper and produce a strongly acidic solution.

In an experiment to investigate this reaction, 0.31 g of white phosphorus reacted in excess aqueous copper(II) sulphate giving 1.60 g of metallic copper. ($A_r$: P = 31, Cu = 64)

**a** i) Calculate the number of moles of phosphorus atoms used.

ii) Calculate the number of moles of copper produced.

iii) Hence calculate the number of moles of copper deposited by one mole of phosphorus atoms.

**b** i) State the change in oxidation number of the copper in this reaction.

ii) Calculate the new oxidation number of the phosphorus after the reaction.

iii) In the reaction the phosphorus forms an acid, $HPO_n$. What is the value of $n$?

**c** Now write a balanced equation showing the action of white phosphorus on copper(II) sulphate in the presence of water.

**d** i) In performing this experiment, what practical difficulty would you expect in weighing the piece of phosphorus?

ii) Describe how you would try to overcome this difficulty.

We now consider in more detail some of the compounds of nitrogen and phosphorus that have been mentioned in your study of the elements, where you have already noticed the existence of several different oxidation states. This is most evident for nitrogen, largely due to the variety of oxides, but there is only one chloride of nitrogen, as you see in the next section.

## ■ 2.2 The halides of nitrogen and phosphorus

**OBJECTIVES**

When you have finished this section you should be able to:
■ explain why **nitrogen forms only one chloride** whereas phosphorus forms two;
■ describe and explain the **hydrolysis of $NCl_3$ and $PCl_3$**;
■ describe the **hydrolysis of $PCl_5$** and the chlorides of other elements in Group V.

Read about the chlorides of nitrogen and phosphorus, paying particular attention to their structures, their stability and their reactions with water. You should then be able to do the following exercises.

**EXERCISE 24**

*Answers on page 93*

**Figure 6**
Energy level diagrams for nitrogen and phosphorus.

Study the energy level diagrams in Fig. 6 and answer the questions that follow.

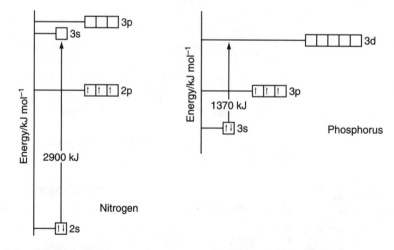

**a** Explain why phosphorus can form two chlorides, $PCl_3$ and $PCl_5$, whereas nitrogen has only one, $NCl_3$.

**b** Phosphorus pentachloride has a molecular structure in the liquid state, but the solid is ionic. Write the formulae of the two ions, each of which contains phosphorus.

**EXERCISE 25**
*Answers on page 93*

a Using your data book, record values of $\Delta H_f^{\ominus}$ for $NCl_3(l)$ and $PCl_3(l)$.
b Write the relevant thermochemical equations for (a).
c What do the values in (a) suggest about the stability of these compounds? Is this borne out by their known properties?

In ILPAC 4, The Periodic Table, you learned that the trichlorides, $NCl_3$ and $PCl_3$, are hydrolysed by water. In the next exercise you remind yourselves of these reactions and extend your knowledge to the hydrolysis of the pentachloride, $PCl_5$.

**EXERCISE 26**
*Answers on page 93*

Write equations and name the products for the reactions of $NCl_3$, $PCl_3$ and $PCl_5$ with water.

The different products obtained when $NCl_3$ and $PCl_3$ are hydrolysed suggest that different mechanisms operate in these reactions. You explore this idea in the next exercise.

**EXERCISE 27**
*Answers on page 94*

The first step in the hydrolysis of phosphorus trichloride is the substitution of an —OH group for —Cl.
a Suggest a mechanism for this substitution (addition/elimination) via the intermediate:

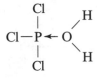

b Two similar steps follow, in which the other two chlorine atoms are replaced by hydroxy groups. Show how the product of these two steps, $P(OH)_3$, rearranges to give phosphonic acid (phosphorous acid).
c Why can the hydrolysis of nitrogen trichloride **not** proceed by a similar mechanism?
d Insert arrows showing electron shifts to complete the following suggested mechanism for the first step in the hydrolysis of nitrogen trichloride.

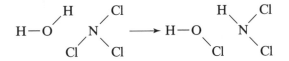

Two similar steps follow, giving $NH_3$ and $3HClO$ as the final products.

As Group V is descended, hydrolysis of the trichloride becomes less complete. In the next exercise you investigate the hydrolysis of antimony and bismuth trichlorides.

**EXERCISE 28**
*Answers on page 94*

The hydrolysis of antimony trichloride is given by the following equation.

$$SbCl_3(aq) + H_2O(l) \rightleftharpoons SbClO(s) + 2HCl(aq)$$
white ppt.
antimony(III) chloride oxide

a What would you observe if concentrated hydrochloric acid is added to the equilibrium mixture above?
b How would you expect bismuth trichloride to react with water? Give a balanced equation.
c Why do laboratory solutions of antimony and bismuth trichlorides contain added hydrochloric acid?

Apart from the trihalides and pentahalides mentioned above, nitrogen forms a fluoride where the nitrogen is in an oxidation state of +1. The compound dinitrogen difluoride, $N_2F_2$, exhibits an interesting form of isomerism which we consider in the next exercise.

**EXERCISE 29**

*Answers on page 94*

A

**a** Draw a dot-and-cross diagram for the molecule $N_2F_2$. (Hint: it contains a double bond.)

**b** What type of isomerism is exhibited by $N_2F_2$? Draw diagrams showing the arrangement of the atoms in the isomers. Name these isomers and indicate briefly what structural feature is responsible for their existence.

Now we turn to another group of compounds, in which different oxidation states are found.

## ■ 2.3 The oxides of nitrogen and phosphorus

**OBJECTIVES**

When you have finished this section you should be able to:
■ place the **common oxides** of:
  **a** nitrogen,
  **b** phosphorus
  in order of increasing **stability**;
■ outline the **preparation of NO, $NO_2$, $N_2O_4$, $P_4O_6$ and $P_4O_{10}$**;
■ explain the meaning of the term **odd-electron molecule** and draw dot-and-cross diagrams for NO, $NO_2$ and $N_2O_4$;
■ draw the **structures of $P_4O_6$ and $P_4O_{10}$**;
■ describe the **action of water on the oxides of nitrogen and phosphorus**.

Table 4, which is incomplete, summarises some information about the oxides of nitrogen and phosphorus. You complete it in the next exercise.

**Table 4**

| Oxidation state of N or P | | +1 | +2 | +3 | +4 | +5 |
|---|---|---|---|---|---|---|
| Nitrogen oxides | Formula | $N_2O$ | | | $NO_2$ ⇌ $N_2O_4$ | |
| | Common name | Nitrous oxide | Nitric oxide | Nitrogen trioxide | | |
| | Recommended name | Dinitrogen oxide | | Dinitrogen trioxide | | |
| | $\Delta H_f^{\ominus}(\Delta G_f^{\ominus})$ /kJ mol$^{-1}$ | | | +84 (+139) | | |
| Phosphorus oxides | Formula | | | $P_4O_6$ | $P_4O_8$ | $P_4O_{10}$ |
| | Common name | | | | Rarely met | |
| | Recommended name | | | | Phosphorus(IV) oxide | |
| | $\Delta H_f^{\ominus}(\Delta G_f^{\ominus})$ /kJ mol$^{-1}$ | | | −1640 (−) | | |

**EXERCISE 30**
*Answers on page 95*

**a** Complete a larger copy of Table 4.
**b** Place the oxides of nitrogen in order of increasing thermal stability under standard conditions.
**c** In view of the fact that the common oxidation states of Group V are +3 and +5, is your answer to (b) surprising?
**d** Of the oxides of nitrogen that do not decompose spontaneously, which is the most likely to relight a glowing splint?
**e** Which of the common oxides of phosphorus is the more stable?

The oxides of nitrogen, given the general formula $NO_x$ are produced in small quantities by direct combustion in the internal combustion engine and cause air pollution. We deal with this in Chapter 6 of ILPAC 5, Introduction to Organic Chemistry. Dinitrogen oxide, $N_2O$, once called 'laughing gas', is used as an anaesthetic in dentistry and as a propellant in whipped cream aerosols.

Dinitrogen oxide in use as an anaesthetic.

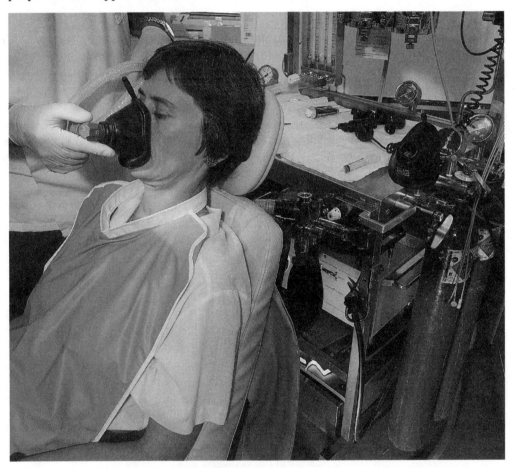

Read about the appearance, structure and laboratory preparation of $NO$, $NO_2$, $N_2O_4$, $P_4O_6$ and $P_4O_{10}$ so that you can do the following two exercises.

**EXERCISE 31**
*Answers on page 96*

**a** Describe the appearance of the following:
   i)   NO at room temperature,
   ii)  $NO_2$ at about 30 °C
   iii) $N_2O_4$ at about 5 °C.
**b** Describe briefly, giving reagents, reaction conditions and equations, how $NO$, $NO_2$ and $N_2O_4$ could be conveniently prepared from nitric acid.
   Comment on the method of collection of NO.
**c** Describe and explain what happens to $N_2O_4$ when it is heated from −10 °C to 600 °C.

**EXERCISE 32**
*Answers on page 97*

a Briefly outline the preparation of $P_4O_6$ and $P_4O_{10}$ from white phosphorus.
b Sketch the structures of $P_4O_6$ and $P_4O_{10}$ from Fig. 7 below.

**Figure 7**

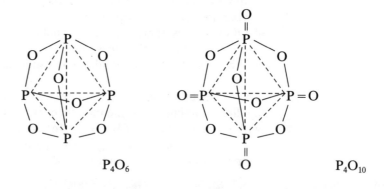

$P_4O_6$          $P_4O_{10}$

How are the structures of the molecules $P_4$, $P_4O_6$ and $P_4O_{10}$ related to each other?
c Why could nitrogen **not** form an oxide with similar structure to $P_4O_{10}$?
d Phosphorus(III) oxide catches fire on heating in air. What product is formed? Write an equation.

The structures of the oxides of phosphorus, as you have seen, can be represented simply by the allocation of outer-shell electrons to bonding pairs and non-bonding pairs (lone pairs). Molecules of some of the oxides of nitrogen, however, have an odd number of outer-shell electrons, at least one of which remains unpaired. For this reason, they are often referred to as 'odd-electron' molecules.

In the next exercise you investigate which of the nitrogen oxides contain unpaired electrons.

**EXERCISE 33**
*Answers on page 97*

With the aid of diagrams, give the electronic structures of NO, $NO_2$ and $N_2O_4$. Comment on any unusual features.

Odd-electron molecules tend to react in a way that will remove the unpaired electron, either by losing it (oxidation) or by pairing it with an unpaired electron in another molecule. With this in mind, attempt the next exercise.

**EXERCISE 34**
*Answers on page 98*

a Why does $NO_2$ form a dimer?
b Would you expect NO to form a dimer? Explain your answer.
c How would you expect NO to react with chlorine?
d You have encountered ions with unpaired electrons in ILPAC 11, Transition Elements. What properties associated with these ions might you also expect for odd-electron molecules?
e What cations might be formed from NO and $NO_2$?

We now go on to consider the reactions of the oxides with water. You may need your textbook to help you with the exercises.

**EXERCISE 35**
*Answers on page 98*

a Although dinitrogen trioxide and dinitrogen pentoxide are not commonly met, they are the anhydrides of well-known acids. Write equations for the reactions of these oxides with water and name the acids.
b Nitrogen dioxide dissolves in water to give a mixture of two acids. Write an equation.
c Nitric acid is manufactured by dissolving nitrogen dioxide in water in the presence of additional oxygen (details in Part B). Write an equation.

**EXERCISE 36**
*Answers on page 98*

**a** Write chemical equations for the reactions of the oxides of phosphorus with water.
**b** Which of the oxides $P_4O_6$ or $P_4O_{10}$ is hydrolysed more readily?
**c** State two laboratory uses of phosphorus(V) oxide.
**d** Predict the change in acid–base character of the oxides in the +3 oxidation state as the group is descended. Explain your answer.

The oxides of arsenic and antimony in the +5 oxidation state are known but they are unstable with respect to the +3 states. Both of these oxides are acidic. Bismuth(V) oxide has never been obtained in the pure form.

You have seen that the oxides of phosphorus, and some of those of nitrogen, are the anhydrides of some well-known acids. We now consider those acids in more detail.

## ■ 2.4 The oxoacids of nitrogen and their salts

Although nitrogen forms six oxides, it forms only two important oxoacids; nitrous acid, $HNO_2$, and nitric acid, $HNO_3$, which you will have come across before.

**OBJECTIVES**

When you have finished this section you should be able to:
■ give examples of **oxidising reactions** of **nitric acid** and **nitrous acid**;
■ identify reactions in which **nitrous acid** and **nitrites** behave as **reducing agents**;
■ explain the unstable nature of nitrous acid.

Read about the chemical properties of nitric acid and nitrous acid in your textbook(s) to obtain an overall picture of these acids.

The next exercise is an A-level question which asks you to make predictions about nitrous acid, $HNO_2$.

**EXERCISE 37**
*Answers on page 99*

The information in this table may be used to predict the likely course of some reactions of nitrous acid, $HNO_2$:

| Half-reaction | | $E^{\ominus}/V$ |
|---|---|---|
| $I_2(aq) + 2e^-$ | $\rightleftharpoons 2I^-(aq)$ | +0.54 |
| $NO_3^-(aq) + 3H^+(aq) + 2e^-$ | $\rightleftharpoons HNO_2(aq) + H_2O(l)$ | +0.94 |
| $HNO_2(aq) + H^+(aq) + e^-$ | $\rightleftharpoons NO(g) + H_2O(l)$ | +0.99 |
| $Br_2(aq) + 2e^-$ | $\rightleftharpoons 2Br^-(aq)$ | +1.09 |

Nitrous acid is normally made freshly when it is required by adding ice-cold aqueous potassium nitrite ($KNO_2$) to dilute hydrochloric acid.

**a** i) Explain why you would not expect nitrous acid to be stable.
  ii) Explain what the effect of a decrease of pH would be on the stability of nitrous acid.
**b** What are the likely products of reaction between nitrous acid and a solution containing iodine and iodide ions?
**c** What are the likely products of reaction between nitrous acid and a solution containing bromine and bromide ions?
**d** State the type of reactant that nitrous acid is in the reactions mentioned in (b) and (c).
**e** i) State the likely products of the reaction between nitrous acid and a solution containing bromide ions and iodide ions.
  ii) What would you expect to **see** in the organic layer if the resulting mixture were shaken with a few drops of cyclohexane (or other suitable organic solvent)?

**EXERCISE 38**

*Answers on page 99*

**a** Calculate the oxidation number of nitrogen in each of the following compounds:

$$HNO_3 \qquad HNO_2 \qquad NO_3^- \qquad NO_2^-$$

**b** Do you think that nitric acid, nitrous acid and their salts could behave as both oxidising and reducing agents?

**c** i) Write a series of common nitrogen compounds in order of decreasing oxidation number to show the products you might expect in redox reactions involving nitric acid, nitrous acid and their salts.

ii) How could you check whether specific predictions were sensible before testing them by experiment?

In the following experiment you test some of the predictions you made in the preceding exercises.

**EXPERIMENT 2** **Reactions of the oxoacids of nitrogen and their salts**

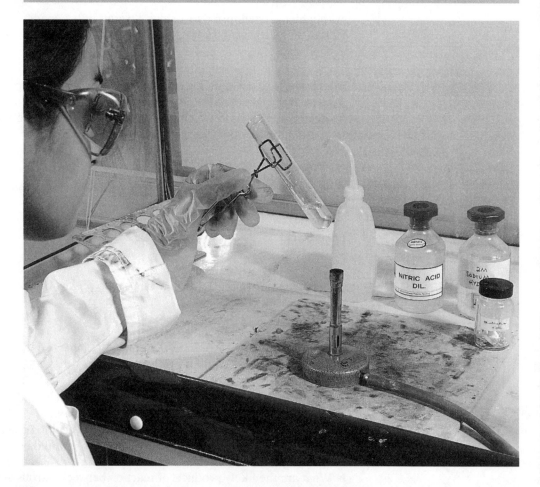

**Aim** The purpose of this experiment is to illustrate some of the redox reactions of nitric acid, nitrates, nitrous acid and nitrites by means of simple small-scale tests.

**Requirements**
- safety spectacles
- protective gloves
- 6 test-tubes
- 1 cork or bung to fit the test-tubes
- copper turnings, Cu

- nitric acid, concentrated, 16 M HNO$_3$
- wash-bottle of distilled water
- Bunsen burner and bench protection mat
- magnesium ribbon, Mg
- nitric acid, dilute, 2 M HNO$_3$
- wood splints
- 2 boiling tubes
- sodium hydroxide solution, 2 M NaOH
- Devarda's alloy (Cu 50%, Al 45%, Zn 5%)
- aluminium foil, Al
- sulphuric acid, dilute, 1 M H$_2$SO$_4$
- beaker, 250 cm$^3$
- ice
- spatula
- sodium nitrite, NaNO$_2$
- potassium iodide solution, 0.5 M KI
- potassium manganate(VII) solution, 0.02 M KMnO$_4$
- iron(II) sulphate, FeSO$_4 \cdot 7H_2O$

## HAZARD WARNING

Concentrated nitric acid is corrosive.

Nitrogen dioxide is a toxic gas, which may be produced in reactions of nitrogen compounds and is a powerful oxidising agent.

Sodium hydroxide solution is corrosive, even in dilute solution.
Therefore you **must**:
- **wear safety spectacles and protective gloves;**
- **work at a fume cupboard.**

**Procedure**
**– Part A**

### Reactions of nitric acid

1. Place a single copper turning in a test-tube and carefully add a few drops of concentrated nitric acid. Record your observations in a copy of Results Table 3 and identify the gas evolved.
2. Pour about 2 cm$^3$ of concentrated nitric acid into a test-tube and carefully dilute it with an equal volume of distilled water. Drop in 2–3 copper turnings. Identify the gas evolved by observing its colour over the whole length of the test-tube. You may need to warm the tube **gently** to speed up the reaction.
3. Add about 4 cm$^3$ of 2 M nitric acid to a 4 cm piece of magnesium ribbon in a test-tube. Cork the tube **loosely** and attempt to identify the gas or gases evolved.
4. Add about 3 cm$^3$ of distilled water to a 4 cm piece of magnesium ribbon in a test-tube and add 1 cm$^3$ of 2 M nitric acid. Cork the tube **loosely** and attempt to identify the gas evolved.
5. In a boiling tube, mix 1 cm$^3$ of 2 M nitric acid and 2 cm$^3$ of 2 M sodium hydroxide solution. Add a spatula measure of Devarda's alloy and warm **gently**. Identify the gas given off.
6. Repeat step 5 using a small piece of aluminium foil instead of Devarda's alloy. Complete your copy of Results Table 3.

**Results Table 3**
Reactions of nitric acid and nitrates

| Reactants | | Observations | Identity of gas |
|---|---|---|---|
| 1. Copper and | 16 M $HNO_3$ | | |
| 2. Copper and | 8 M $HNO_3$ | | |
| 3. Magnesium and | 2 M $HNO_3$ | | |
| 4. Magnesium and | 0.5 M $HNO_3$ | | |
| 5. Devarda's alloy and a nitrate in alkali | | | |
| 6. Aluminium and a nitrate in alkali | | | |

**– Part B**  **Reactions of nitrous acid**

7. Prepare a solution of nitrous acid, as follows. Pour about 15 cm$^3$ of dilute sulphuric acid into a boiling tube and stand it in an ice-bath for about 5 minutes. Dissolve about 1.5 g of sodium nitrite in the minimum quantity of distilled water and cool in the ice-bath. Mix the two cooled solutions.

8. Transfer about 2 cm$^3$ of the nitrous acid solution to a test-tube and warm gently. Record your observations and try to identify the gas evolved.

9. In another test-tube add a few drops of potassium iodide solution to approximately 2 cm$^3$ of nitrous acid solution.

10. Repeat step 9 using a few drops of potassium manganate(VII) solution instead of potassium iodide.

11. Dissolve a few small crystals of iron(II) sulphate in about 2 cm$^3$ of cold distilled water.

    **a** To half of this solution add dilute aqueous sodium hydroxide.

    **b** Add the other half of the iron(II) sulphate solution to an equal volume of nitrous acid solution. Heat the mixture until the colour lightens and then cool. Add sodium hydroxide solution and compare the result with what happened in (a).

12. To a little solid sodium nitrite add 1–2 cm$^3$ of aqueous sodium hydroxide and a small piece of aluminium foil. Heat the mixture and test any gases evolved. Complete your copy of Results Table 4.

**Results Table 4**
Reactions of nitrous acid and nitrites

| Reactants in solution | Observations | Identity of coloured product(s) | Oxidation or reduction of $NO_2^-$ |
|---|---|---|---|
| 7. Sodium nitrite and sulphuric acid | | $HNO_2$ or perhaps $N_2O_3$ | |
| 8. Warm nitrous acid | | | |
| 9. Potassium iodide and nitrous acid | | | |
| 10. Potassium manganate(VII) and nitrous acid | | | |
| 11a. Iron(II) sulphate and sodium hydroxide | | | |
| 11b. Iron(II) sulphate, nitrous acid and sodium hydroxide | | | |
| 12. Aluminium and a nitrite in alkali | | | |

*Specimen results on page 101*

**Questions**

*Answers on page 101*

To help you with your answers you may need to refer to your textbooks or to Exercises 37 and 38.
1. Write equations for the reactions between copper and nitric acid of different concentrations. Explain your observations of these reactions.
2. Describe and explain the difference between the reactions of magnesium with dilute and very dilute nitric acid.
3. Aluminium is the most powerful reducing agent of the three metals in Devarda's alloy. Suggest reasons why the alloy is more effective than aluminium foil.
4. Write an equation for the thermal decomposition of nitrous acid. What type of redox reaction is this? Why does nitric acid not undergo the same type of reaction?
5. Which of the reactions in the experiment may be used to distinguish:
   **a** nitrates and nitrites from most other compounds,
   **b** nitrates from nitrites?

The next exercise tests your understanding of the redox reactions of nitric acid.

**EXERCISE 39**
*Answers on page 102*

Under certain concentration and temperature conditions, 0.8 g of iron is found to react with 0.9 g of pure nitric acid, evolving an oxide of nitrogen ($N_2O$, or $NO$, or $N_2O_4$).
**a** Calculate the molar ratio of iron and nitric acid that react.
**b** Complete:

$$\text{Gain of} \ldots \text{electrons}$$

$$\ldots \text{Fe(s)} + \ldots \text{HNO}_3\text{(aq)} \longrightarrow \ldots \text{Fe}^{3+}\text{(aq)} + \begin{bmatrix} N_2O \\ \text{or } NO \\ \text{or } N_2O_4 \end{bmatrix}$$

$$\text{Loss of} \ldots \text{electrons}$$

**c** From the number of electrons gained by the nitric acid, deduce the change in oxidation number of nitrogen and decide which of the three oxides of nitrogen is formed.
**d** Complete:

$$\ldots \text{Fe(s)} - \ldots \text{e}^- \rightarrow \ldots \text{Fe}^{3+}\text{(aq)}$$

$$\ldots \text{H}^+ + 1\text{HNO}_3\text{(aq)} + \ldots \text{e}^- \rightarrow \begin{bmatrix} \frac{1}{2}N_2O \\ \text{or } 1NO \\ \text{or } \frac{1}{2}N_2O_4 \end{bmatrix} + \ldots \text{H}_2O$$

and so, by addition, balance the equation for the reaction between iron and nitric acid under these conditions. ($H = 1$, $N = 14$, $O = 16$, $Fe = 56$)

We deal with the industrial production and uses of nitric acid and nitrogenous fertilisers in Part B of this book.
Now we consider the most important of the oxoacids of phosphorus – phosphoric(V) acid – and some of its salts.

## ■ 2.5 Phosphoric(V) acid and phosphates

Phosphorus forms fewer oxides than nitrogen, but many more oxoacids. However, we deal in this section only with the commonest oxoacid of phosphorus, phosphoric(V) acid, commonly called orthophosphoric acid. This is a colourless solid when pure, but is more often seen as a syrupy liquid containing a small amount of water.

**OBJECTIVES**
When you have finished this section you should be able to:
■ draw the **shapes** of $H_3PO_4$, $PO_4^{3-}$, $HPO_4^{2-}$ and $H_2PO_4^-$;
■ state some **uses of phosphoric(V) acid and phosphates**.

In the next exercise you build on what you have learned about phosphoric(V) acid in the other ILPAC books, but you may also need to refer to your textbook(s).

**EXERCISE 40**
*Answers on page 102*

**a** Write three equations to show the reactions that occur when an alkali is added to a solution of phosphoric(V) acid, $H_3PO_4$. Name the ions containing phosphorus.
**b** Would you expect a solution of trisodium phosphate, $Na_3PO_4$, to be acidic, alkaline or neutral? Explain briefly.
**c** The structural formula of phosphoric(V) acid can be represented as $PO(OH)_3$. What shape(s) would you expect for this molecule and for the ions formed from it?
**d** Why could nitric(V) acid not have a similar structure?
**e** What type of intermolecular bonding would you expect to be predominant in phosphoric(V) acid in its solid and 'syrupy' states?
**f** Why is phosphoric(V) acid used in preference to sulphuric acid for the preparation of hydrogen bromide from a salt such as potassium bromide? Write an equation.
**g** Why is phosphoric(V) acid sometimes used in preference to sulphuric acid for the dehydration of organic molecules?

To help you with the following two exercises, read about the uses of phosphates, particularly in fertilisers and water softening.

**EXERCISE 41**

*Answers on page 103*

Phosphorus is essential for plant root growth and is one of the three most important elements (N, P, K) in artificial fertilisers. Compounds known as 'super-phosphate' and 'triple-phosphate' are made from calcium phosphate(V), $Ca_3(PO_4)_2$, which is found naturally as 'rock phosphate'.

**a** Why is calcium phosphate not often used as a fertiliser?

**b** How is calcium phosphate converted to:
   i) 'superphosphate',
   ii) 'triple-phosphate' (triple-superphosphate)?

**c** What is the difference in chemical constitution between the superphosphate and the triple-phosphate?

**d** Some phosphate ions (e.g. the polyphosphate ion, $P_6O_{18}^{6-}$) form fairly stable complex ions (see ILPAC 11, Group IV Elements and Transition Elements) with calcium and magnesium ions. Explain how the addition of sodium polyphosphate, $Na_6P_6O_{18}$, (known commercially as 'Calgon') softens hard water. The use of such softeners is now controlled in some countries because of its effect on river water (in other countries treatment of sewage outflow to remove phosphates is compulsory) – why do you think this is necessary?

Spreading phosphate fertiliser.

We now turn our attention from the oxoacids of Group V elements to the hydrides.

## ■ 2.6 The hydrides of nitrogen and phosphorus

**OBJECTIVES**

When you have finished this section you should be able to:
- list the **hydrides** formed by Group V elements;
- compare the **shapes of the molecules** of **ammonia** and **phosphine**;
- compare the **boiling points** and the **solubilities** in water of ammonia and phosphine;
- state and explain the trend in the **thermal stability** of the trihydrides as the group is descended;
- explain the **difference in base strength** between ammonia and phosphine in aqueous solution;
- describe the effect of heat and of water on **phosphonium iodide**;
- compare the **reducing and complexing properties** of ammonia and phosphine.

All the Group V elements form trihydrides, $XH_3$. The stability of these hydrides falls rapidly down the group, so that $SbH_3$ and $BiH_3$ are very unstable, the latter having been obtained only in traces. Read about these hydrides to help you with the following exercises.

**EXERCISE 42**

*Answers on page 103*

Study Table 5 and answer the following questions about the hydrides of Group V.

**Table 5**

| Element | Name and formula of hydride, $XH_3$ | Other hydrides include |
|---|---|---|
| Nitrogen<br>Phosphorus<br>Arsenic<br>Antimony<br>Bismuth | Ammonia, $NH_3(g)$<br>Phosphine, $PH_3(g)$<br>Arsine, $AsH_3(g)$<br>Stibine, $SbH_3(g)$<br>Bismuthine, $BiH_3(g)$ | Hydrazine, $N_2H_4(l)$<br>Diphosphane, $P_2H_4(l)$ |

**a** What are the oxidation numbers of the Group V elements in the hydrides listed in Table 5?

**b** Predict the shape of the $PH_3$ molecule from your knowledge of the shape of the $NH_3$ molecule.

**c** The bond angles in $PH_3$ are in fact 93°, much smaller than the 107° in $NH_3$. This difference is not easy to explain in terms of electron pair repulsion. A better explanation for the shape of the $PH_3$ molecule is that $sp^3$ hybridisation does not occur as it does in $NH_3$, i.e. the lone pair is in the 3s orbital and the bonding pairs are in 3p orbitals. How does this account for the difference in bond angles?

**EXERCISE 43**

*Answers on page 104*

**a** Write down the boiling points of the first four hydrides, $XH_3$, in Group V. Use your data book.

**b** Comment on the trend in (a) and state the exception to the trend.

**c** Why is the boiling point of ammonia higher than that of phosphine even though the relative molecular mass of phosphine is greater? (Hint: consider the electronegativities of N, P and H.)

**d** Why do you think ammonia is soluble in water whereas phosphine is virtually insoluble?

In the following exercise you consider the thermal stability of the $XH_3$ hydrides.

**EXERCISE 44**

*Answers on page 104*

**a** Ammonia is reasonably stable to heat, phosphine and arsine both decompose on heating and stibine and bismuthine are both unstable at room temperature.
   Write equations to show the thermal decomposition of ammonia and phosphine.

**b** Complete a copy of Table 6 with the aid of your data book.

**Table 6**

| Bond | Mean bond energy/kJ mol$^{-1}$ |
|---|---|
| N—H<br>P—H<br>As—H<br>Sb—H<br>Bi—H | <br><br>292<br>255<br>Not listed |

**c** Do you think the bond energy terms listed in Table 6 are sufficient to explain the trend in thermal stability?

The presence of arsenic in food was once detected in forensic laboratories by the decomposition of arsine produced from the action of zinc and dilute acid on arsenic compounds. (See Fig. 8.)

**Figure 8**
An illustration of an old apparatus used to detect arsenic in food.

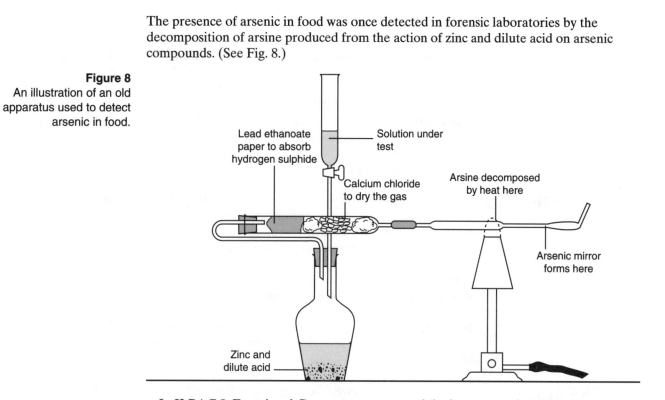

In ILPAC 8, Functional Groups, you compared the base strengths of ammonia and primary amines by considering the availability of the lone pairs of electrons on the nitrogen atoms and the relative stabilities of $NH_4^+$ and $RNH_3^+$. In the next exercise you similarly compare the base strength of ammonia with that of phosphine.

**EXERCISE 45**
*Answers on page 104*

**a** Ammonia dissociates in water as follows:

$$NH_3(aq) + H_2O(l) \rightleftharpoons NH_4^+(aq) + OH^-(aq)$$
$$\text{or } NH_3{\cdot}H_2O(aq) \rightleftharpoons NH_4^+(aq) + OH^-(aq)$$

The base dissociation constant, $K_b$(298 K), for this reaction is $1.77 \times 10^{-5}$ mol dm$^{-3}$.
   For a similar reaction involving the small amount of phosphine that dissolves, $K_b$ is about $10^{-26}$ mol dm$^{-3}$.
   Which is the more basic, phosphine or ammonia?

**b** Suggest a reason for the difference in base strength with reference to the orbital structures shown in Fig. 9 below:

**Figure 9**

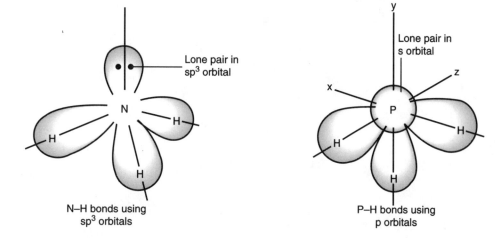

N–H bonds using sp³ orbitals

P–H bonds using p orbitals

**c** Would you expect the phosphonium ion, $PH_4^+$, to be more or less stable than $NH_4^+$? (Hint: look at your answer to (b) and at your copy of Table 6 in Exercise 44.)

**d** Would you expect phosphine to form complex ions with transitional metal ions similar to, say, $Cu(NH_3)_4^{2+}$?

Since water is a stronger base than phosphine, you might expect reactions to occur between phosphonium compounds and water similar to those that occur between ammonium compounds and dilute alkalis.

From your knowledge of ammonium compounds, attempt the next exercise, which deals with phosphine and phosphonium ($PH_4^+$) compounds.

**EXERCISE 46**
*Answers on page 104*

**a** Write an equation for the reaction between phosphine and hydrogen chloride.

**b** Write an equation to show the effect of heat on phosphonium iodide.

**c** All phosphonium salts are immediately hydrolysed by water and dilute alkalis. Give two equations for this decomposition, one with water and the other with dilute sodium hydroxide.

**d** How do the reactions in (c) differ from the reactions with ammonium compounds?

**e** What is the shape of the phosphonium ion?

Now you compare the reducing powers of ammonia and phosphine by considering their reactions with oxygen and chlorine. Read about the reducing properties of the hydrides for the next exercise.

**EXERCISE 47**
*Answers on page 105*

**a** Phosphine is a stronger reducing agent than ammonia. Illustrate this fact with reference to the reaction of the two hydrides with oxygen and chlorine.

**b** Calculations of $\Delta H^{\ominus}$ (and $\Delta G^{\ominus}$) for the combustion of ammonia and phosphine give large negative values in each case, which indicates that ammonia might react more readily than it does. Suggest reasons for its lack of reactivity in this context.

**c** How does ammonia react with copper(II) oxide?

The flammability of phosphine in air is supposed to be responsible for the flickering lights ('Will o' the Wisp' or 'Jack o' Lanterns') observed near marshes, caves and tombs. Small traces of impure phosphine are produced by bacterial decay of organic matter, the impurities causing it to be spontaneously flammable.

To consolidate your knowledge of the chemistry of nitrogen and phosphorus, attempt the following teacher-marked exercise.

**EXERCISE**
*Teacher-marked*

Describe and account for the similarities and differences in the chemistry of nitrogen and phosphorus, using:

**a** the elements,

**b** the hydrides,

**c** the chlorides as examples.

We deal with the industrial production of ammonia in Part B. Now we turn our attention to the most important elements in the next group of the Periodic Table.

# CHAPTER 3

# GROUP VI: OXYGEN AND SULPHUR

The Group VI elements show the usual gradation from non-metallic to metallic properties with increasing atomic number. Oxygen and sulphur are non-metals, selenium and tellurium are semiconductors and polonium is metallic and radioactive.

As with Groups III and V, we consider the first two members of the group in more detail.

## ■ 3.1 The nature of the elements

Oxygen and sulphur are widespread in nature, both as elements and in compounds, and they exhibit allotropy (polymorphism).

**OBJECTIVES**

When you have finished this section you should be able to:
■ describe the **appearance and structures** of the **allotropes of oxygen** and **sulphur**;
■ explain why oxygen is a gas while sulphur is a solid;
■ state the reactions of oxygen and sulphur with **selected metals** and **non-metals**;
■ list the common **oxidation states** of oxygen and sulphur.

To get a simple overall picture of Group VI, read the general introduction to the group in your textbook(s). Then look in more detail at the allotropes of oxygen and sulphur paying particular attention to their molecular structures. Also read about the physical and chemical properties of these elements.

In the next exercise you consider the allotropes of oxygen and sulphur, and investigate the similarities that exist between the elements.

**EXERCISE 48**
*Answers on page 105*

**a** Name and describe the appearance of the allotropes of sulphur and oxygen.
**b** Draw the shapes of the molecules for the allotropes named in (a). (Indicate any multiple bonds.)
**c** Between 96° and its melting point at 119° monoclinic sulphur is more stable than rhombic sulphur, while below 96° the reverse is true. 96° is known as the transition temperature, at which the two forms are at equilibrium, although the change from one form to the other is very slow. Given a suitable solvent or solvents, suggest how you could make crystals of the two forms from powdered sulphur.
**d** Write equations for the reactions of oxygen and sulphur with the following elements, and name the products.
  i)   sodium,
  ii)  iron,
  iii) hydrogen,
  iv) carbon.

Apart from the similarities in the reactions with the selected metals and non-metals and the existence of allotropes, oxygen and sulphur have very little else in common. The immediately striking difference is that oxygen is a colourless gas while sulphur is a yellow crystalline solid.

Polarised light micrograph of crystals of sulphur.

In the last exercise you learned that oxygen tends to form double bonds with other oxygen atoms producing diatomic molecules, whereas sulphur tends to catenate and form $S_8$ ring molecules. In the next exercise you investigate the reason for this.

## EXERCISE 49

*Answers on page 106*

Use the mean bond energies, $\overline{E}$, shown in Table 7 to answer the questions that follow.

**Table 7**

| Bond | O=O | O—O | S=S | S—S |
|---|---|---|---|---|
| $\overline{E}$/kJ mol$^{-1}$ | 498 | 142 | 431 | 264 |

**a** Calculate the enthalpies of reaction for the following changes. Assume that $S_2$ and $O_8$ have the same structures as $O_2$ and $S_8$.
  i)  $S_8(g) \rightarrow 4S_2(g)$
  ii) $O_8(g) \rightarrow 4O_2(g)$

**b** Why does oxygen consist of diatomic molecules whereas sulphur contains $S_8$ molecules?

In the next exercise you look at the electronic configurations of isolated atoms of oxygen and sulphur and explain how the possible oxidation states of the elements arise.

## EXERCISE 50

*Answers on page 106*

**Figure 10**
Electron configurations of oxygen and sulphur.

**a** Complete a copy of Fig. 10 (not drawn to scale).

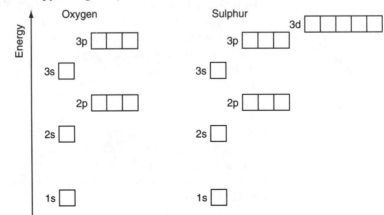

**b** Why are both elements commonly divalent?

**c** What is the apparent covalency of oxygen in $H_3O^+$? Explain how this occurs.

**d** As well as forming the $S^{2-}$ ion, sulphur can form two, four and six covalent bonds with other atoms (e.g. in $H_2S$, $SCl_2$, $SO_2$ and $SF_6$). Explain how this occurs.

**e** Why is oxygen unable to form four and six covalent bonds with other atoms?

**f** The +6 oxidation state is very common for S, Se and Te but the +4 oxidation state is the most important for Po. What name do we give this effect and where have you met it before?

We now go on to consider some important compounds of oxygen and sulphur. Since oxygen is usually limited to the −2 oxidation state, there are many compounds of sulphur that have no oxygen analogues, e.g. $SCl_4$ and $SF_6$ exist but $OCl_4$ and $OF_6$ are not known.

The first class of compound we consider is the hydrides.

# ■ 3.2 The hydrides of oxygen and sulphur

The hydrides $H_2O$, $H_2S$, $H_2Se$ and $H_2Te$ are all known and, with the notable exception of water, are all poisonous and pungent gases. Another anomalous property of oxygen is that it also forms another type of hydride, $H_2O_2$.

**OBJECTIVES**

When you have finished this section you should be able to:
■ describe and, as far as possible, explain the **differences between water and hydrogen sulphide** in respect of:
   **a** boiling points,
   **b** shapes of molecules,
   **c** thermal stability,
   **d** reducing action,
   **e** acidity;
■ explain how hydrogen sulphide causes **precipitation of metal sulphides**;
■ describe the **physical properties of hydrogen peroxide** and explain its **chemical instability**;
■ quote some examples of **redox reactions** involving hydrogen peroxide.

Read about the properties of water and hydrogen sulphide so that you can attempt the following exercises. You have, of course, already done some work on them in other ILPAC books.

**EXERCISE 51**
*Answers on page 107*

The boiling points of water, $H_2O$, and hydrogen sulphide, $H_2S$, are 373 K and 212 K respectively. The relative molecular masses of water and hydrogen sulphide are 18 and 34 respectively.

**a** By referring to the relative molecular masses, state which of the two substances you would expect to have the higher boiling point and give a reason.

**b** Explain the differences in boiling points in terms of intermolecular bonding.

**EXERCISE 52**
*Answer on page 107*

Explain why the structure of hydrogen sulphide is not linear.

You are probably already aware of the high thermal stability of water. Even at 2000 °C it is only slightly decomposed into its elements. Hydrogen sulphide, however, is decomposed by strong heating. The next exercise concerns the thermal stabilities of $H_2O$ and $H_2S$ and the ease with which these compounds are formed from their elements.

**EXERCISE 53**

*Answers on page 108*

**a** Look up values of $\Delta H_f^{\ominus}$ and $\Delta G_f^{\ominus}$ for $H_2O(g)$ and $H_2S(g)$ in your data book. Comment on the ease with which these compounds can be made from their elements.
**b** Write down the mean bond energies for O—H and S—H using your data book.
**c** Account for the higher thermal stability of water compared with hydrogen sulphide.
**d** The relative instability of hydrogen sulphide is also shown by the fact that it burns readily in air. Write an equation for this reaction.

Since $H_2S$ gives up its hydrogen atoms easily it is an effective reducing agent, reacting with many oxidising agents, as you see in the next exercise.

**EXERCISE 54**

*Answers on page 108*

**Table 8**

Using the information displayed in Table 8, answer the questions that follow.

| Half-reaction | | $E^{\ominus}$/V |
|---|---|---|
| 1. | $S(s) + 2H^+(aq) + 2e^- \rightleftharpoons H_2S(aq)$ | +0.14 |
| 2. | $O_2(g) + 4H^+(aq) + 4e^- \rightleftharpoons 2H_2O(l)$ | +1.23 |
| 3. | $Fe^{3+}(aq) + e^- \rightleftharpoons Fe^{2+}(aq)$ | +0.77 |
| 4. | $MnO_4^-(aq) + 8H^+(aq) + 5e^- \rightleftharpoons Mn^{2+}(aq) + 4H_2O(l)$ | +1.51 |
| 5. | $SO_2(aq) + 4H^+(aq) + 4e^- \rightleftharpoons S(s) + 2H_2O(l)$ | +0.45 |
| 6. | $SO_4^{2-}(aq) + 4H^+(aq) + 2e^- \rightleftharpoons SO_2(aq) + 2H_2O$ | +0.17 |
| 7. | $\frac{1}{2}Cl_2(aq) + e^- \rightleftharpoons Cl^-(aq)$ | +1.36 |

**a** What do the $E^{\ominus}$ values of half-reactions 1 and 2 suggest about the relative strengths of water and hydrogen sulphide as reducing agents?
**b** Would you expect hydrogen sulphide to reduce the following substances:
   i)   aqueous iron(III) chloride,
   ii)  aqueous potassium manganate(VII),
   iii) sulphur dioxide,
   iv)  sulphuric acid,
   v)   chlorine?
   Explain your answers.
**c** Write balanced equations for the reactions occurring in (b).

One property common to all Group VI hydrides is their ability to function as weak acids. The next three exercises deal with this property.

**EXERCISE 55**

*Answers on page 109*

Ionisation of hydrogen sulphide in water occurs in two stages:

$$H_2O(l) + H_2S(aq) \rightleftharpoons H_3O^+(aq) + HS^-(aq)$$
hydrosulphide ion
$$H_2O(l) + HS^-(aq) \rightleftharpoons H_3O^+(aq) + S^{2-}(aq)$$
sulphide ion

The acid dissociation constants are, respectively, $10^{-7}$ mol dm$^{-3}$ and $10^{-14}$ mol dm$^{-3}$.
**a** Identify the acid and base in each ionisation stage.
**b** Arrange the substances appearing in the equations in order of decreasing concentration for a saturated solution of hydrogen sulphide.

In the above exercise you saw how hydrogen sulphide reacts with a weak base, water. In the next exercise you consider reactions with stronger bases.

**EXERCISE 56**

*Answers on page 109*

**a** What compounds will be formed in solution if hydrogen sulphide reacts with sodium hydroxide solution so that:
   i)  hydrogen sulphide is in excess,
   ii) sodium hydroxide is in excess?
**b** Write equations for the reactions occurring in (a).
**c** Why do you think most s-block sulphides give an alkaline aqueous solution that smells unpleasant? Illustrate your answer with reference to sodium sulphide, $Na_2S$.

The next exercise deals with the precipitation of metal sulphides.

**EXERCISE 57**
*Answers on page 109*

**a** Only a few metals form hydrogensulphides, such as NaHS and KHS, which exist as solids. Suggest a reason for this. (Hint: what factors affect the stability of other solids with large anions?)
**b** Most metals form sulphides, and many of these are very insoluble. With reference to the equations in Exercise 55, explain how bubbling hydrogen sulphide through a solution of a metal salt can cause complete precipitation of the metal sulphide. Use lead(II) sulphide as an example.
**c** Some 'insoluble' sulphides are not precipitated by hydrogen sulphide if the pH of the solution is lowered. Why is this?

In the next exercise you compare the acid strengths of the Group VI hydrides.

**EXERCISE 58**
*Answers on page 110*

Values of $pK_a$ for the first ionisation of the hydrides of Group VI are given below.

| Hydride | $H_2O$ | $H_2S$ | $H_2Se$ | $H_2Te$ |
|---------|--------|--------|---------|---------|
| $pK_a$  | 14     | 7      | 4       | 3       |

**a** How does acid strength change as the group is descended?
**b** Suggest a possible reason for this trend.

Other hydrides can also be formed by elements in this group, the most important being hydrogen peroxide, $H_2O_2$. You learned about the preparation of hydrogen peroxide in ILPAC 4, s-Block Elements; now you consider some of its properties.

Read about the properties of hydrogen peroxide in your textbook(s) so that you can do the following exercises.

**EXERCISE 59**
*Answers on page 110*

**a** Draw the shape of the $H_2O_2$ molecule.
**b** Hydrogen peroxide, in its liquid state or in very concentrated solutions, is dangerously explosive but its enthalpy of formation is $-187$ kJ mol$^{-1}$ ($\Delta G_f^{\ominus} = -118$ kJ mol$^{-1}$). Explain this apparent anomaly.
**c** Pure hydrogen peroxide is a pale blue syrupy liquid boiling at 152 °C. How do you account for its viscous nature and high boiling point?
**d** Would you expect hydrogen peroxide to be acidic?

Hydrogen peroxide is a powerful and useful oxidising agent, but it can also be oxidised, as you see in the next exercise.

**EXERCISE 60**
*Answers on page 110*

**Table 9**

Use the standard electrode potentials given in Table 9 to answer the questions that follow:

| Half-reaction | $E^{\ominus}$/V |
|---------------|------|
| 1.  $S(s) + 2H^+(aq) + 2e^- \rightleftharpoons H_2S(aq)$ | +0.14 |
| 2.  $I_2(aq) + 2e^- \rightleftharpoons 2I^-(aq)$ | +0.54 |
| 3.  $O_2(g) + 2H^+(aq) + 2e^- \rightleftharpoons H_2O_2(aq)$ | +0.68 |
| 4.  $MnO_4^-(aq) + 8H^+(aq) + 5e^- \rightleftharpoons Mn^{2+}(aq) + 4H_2O(l)$ | +1.51 |
| 5.  $H_2O_2(aq) + 2H^+(aq) + 2e^- \rightleftharpoons 2H_2O(l)$ | +1.77 |

**a** Write equations for the reactions in solution between hydrogen peroxide and:
   i)   hydrogen sulphide,
   ii)  iodide ions,
   iii) potassium manganate(VII).
**b** Write an equation for the disproportionation of hydrogen peroxide, showing the oxidation state of oxygen in reactant and products.
**c** What do you think would happen if manganese(II) ions were added to aqueous hydrogen peroxide?
**d** Equation 5 shows a useful feature of the use of hydrogen peroxide as an oxidant (in addition to a high value of $E^{\ominus}$). What is this?

In the teacher-marked exercise that follows, you can consolidate your knowledge of Group VI elements, drawing not only on this book but also on information gained from other ILPAC books.

**EXERCISE**
*Teacher-marked*

**a** Compare the properties of water and hydrogen sulphide, giving explanations of any similarities or differences that you quote.
**b** Comment on the following observations.
  i)  Oxygen normally exists in the form of diatomic molecules, whereas sulphur normally exists in the form of $S_8$ molecules.
  ii) Oxygen does not form a hexafluoride analogous to $SF_6$.
  iii) There are no compounds containing the $O^-$ or $S^-$ ions although the electron affinities of oxygen and sulphur are $-142$ kJ mol$^{-1}$ and $-200$ kJ mol$^{-1}$ respectively.
  iv) Oxygen has a higher electronegativity than sulphur although more energy is released when a sulphur atom accepts an electron than when an oxygen atom accepts an electron.

In the remaining sections of Part A, we deal with the oxides and oxoacids of sulphur. There are, of course, no corresponding compounds of oxygen.

## ■ 3.3 The oxides of sulphur

Sulphur forms only two well-known oxides, sulphur dioxide, $SO_2$, and sulphur trioxide, $SO_3$, both of which are acidic.

**OBJECTIVES**   When you have finished this section you should be able to:
  ■ describe the **structure** of molecules of **sulphur dioxide** and **sulphur trioxide**;
  ■ describe reactions in which **sulphur dioxide acts as a reducing agent**.

Read about the structure, bonding and properties of sulphur dioxide and sulphur trioxide in your textbook(s) to help you with the following exercises. Look for the structure of the $SO_3$ molecule rather than the polymeric solid forms.

**EXERCISE 61**
*Answers on page 111*

**a** Draw dot-and-cross diagrams for $SO_2$ and $SO_3$.
**b** How do you explain the fact that both S—O bond lengths in $SO_2$ are equal and that all three in $SO_3$ are equal?
**c** Describe the shapes of the $SO_2$ and $SO_3$ molecules, indicating the bond angles approximately.

**EXERCISE 62**
*Answers on page 111*
**Table 10**

Use the information in Table 10 to answer the questions that follow.

| Half-reaction | $E^{\ominus}$/V |
|---|---|
| 1.  $Cr_2O_7{}^{2-}(aq) + 14H^+(aq) + 6e^- \rightleftharpoons 2Cr^{3+}(aq) + 7H_2O(l)$ | +1.33 |
| 2.  $Fe^{3+}(aq) + e^- \rightleftharpoons Fe^{2+}(aq)$ | +0.77 |
| 3.  $MnO_4{}^-(aq) + 8H^+(aq) + 5e^- \rightleftharpoons Mn^{2+}(aq) + 4H_2O(l)$ | +1.51 |
| 4.  $SO_2(aq) + 4H^+(aq) + 4e^- \rightleftharpoons S(s) + 2H_2O(l)$ | +0.45 |
| 5.  $SO_4{}^{2-}(aq) + 4H^+(aq) + 2e^- \rightleftharpoons SO_2(aq) + 2H_2O$ | +0.17 |
| 6.  $Cl_2(aq) + 2e^- \rightleftharpoons 2Cl^-(aq)$ | +1.36 |

**a** Would you expect sulphur dioxide to reduce the following reagents?
  i)  $MnO_4{}^-(aq)$,
  ii) $Fe^{3+}(aq)$,
  iii) $Cl_2(aq)$,
  iv) $Cr_2O_7{}^{2-}(aq)$.
**b** Write balanced equations for the reactions which do occur.
**c** Which of the reactions forms the basis of a useful test for gaseous sulphur dioxide?

You know already that the oxides of sulphur are the anhydrides of sulphurous and sulphuric acids. In the next section, you learn more about these oxoacids and some others that you may not have met before.

## ■ 3.4 The oxoacids and oxosalts of sulphur

Sulphur forms many oxoacids, like nitrogen, phosphorus and chlorine, its near neighbours in the Periodic Table. The commonest acid is sulphuric acid, $H_2SO_4$. The other acids are better known as their salts.

**OBJECTIVES**

When you have finished this section you should be able to:
- explain why **sulphuric acid** is a **stronger acid** than **sulphurous acid**;
- state and explain the **physical and chemical properties** of sulphuric acid;
- state and explain the chemical properties of the following anions:
  **a** **sulphite**, $SO_3^{2-}$,
  **b** **sulphate**, $SO_4^{2-}$,
  **c** **thiosulphate**, $S_2O_3^{2-}$,
  **d** **peroxodisulphate**, $S_2O_8^{2-}$ (persulphate);
- describe the **structure and bonding** of the sulphite, sulphate and thiosulphate ions.

An aqueous solution, often called 'sulphurous acid', $H_2SO_3$, is made by passing sulphur dioxide into water. It is slowly oxidised by air to sulphuric acid. The solution contains the ions $SO_3^{2-}$, $HSO_3^-$ and $H_3O^+$ as well as dissolved $SO_2$ molecules, but there is no strong evidence for the molecule $H_2SO_3$. Therefore, sulphurous acid is best represented by the equilibrium shown in the next exercise.

**EXERCISE 63**
*Answers on page 112*

Solutions of sulphur dioxide are believed to contain hydrated molecules of $SO_2(aq)$ which ionise as follows.

$$SO_2(aq) + H_2O(l) \rightleftharpoons H^+(aq) + HSO_3^-(aq) \qquad (1)$$
$$HSO_3^-(aq) \rightleftharpoons H^+(aq) + SO_3^{2-}(aq) \qquad (2)$$

**a** From your data book write down the $K_a$ and $pK_a$ values for the two ionisations shown above.
**b** Name the ions shown in equations *(1)* and *(2)*.
**c** Write equations for the two ionisation stages that occur in dilute sulphuric acid.
**d** From your data book write down the $K_a$ and $pK_a$ values for the two ionisation stages in sulphuric acid.
**e** Compare the acid strengths of sulphuric and 'sulphurous' acids.

Sulphuric acid is a reagent that you have used both in its dilute and concentrated forms throughout this course. In the teacher-marked exercise that follows, you are asked to bring together the information you have gained from various ILPAC books about sulphuric acid.

**EXERCISE**
*Teacher-marked*

Sulphuric acid is one of the most versatile reagents used in the laboratory. Discuss this statement by referring to reactions that illustrate its behaviour:
**a** as an acid,
**b** in the displacement of other acids from their salts,
**c** as a sulphonating agent (if this is part of your organic chemistry syllabus),
**d** as an oxidising agent,
**e** as a dehydrating agent.

Another teacher-marked exercise follows, which includes some chemistry of nitrogen.

**EXERCISE**
*Teacher-marked*

**a** Give an account of the **redox** chemistry of the anions of nitrogen and sulphur. Your answer should include reference to the chemistry of $NO_2^-$, $NO_3^-$, $S^{2-}$, $SO_3^{2-}$, $S_2O_3^{2-}$ and $S_2O_8^{2-}$. You should consider the relative stabilities of the various oxidation states exhibited by nitrogen and sulphur, and include a range of reactions to illustrate your account. Marks will be gained for properly balanced equations.

**b** When a heavy metal nitrate is heated to produce the metal oxide, nitrogen dioxide and oxygen, the nitrogen dioxide and oxygen are always produced in the ratio 4:1. By considering the oxidation state changes of the nitrogen and the oxygen show why this should be so, and hence write an equation for the thermal decomposition of aluminium nitrate.

We deal with the industrial production and uses of sulphuric acid in Part B. Now you investigate, by experiment, the properties of some of the oxosalts of sulphur.

## EXPERIMENT 3 Investigating some reactions of the oxosalts of sulphur

**Aim**

The purpose of this experiment is to illustrate the redox reactions of some oxoanions of sulphur, some of which can be used as tests for these ions.

**Introduction**

You are asked to carry out some simple test-tube reactions on solutions and solid samples of the following salts:
■ sodium sulphite, $Na_2SO_3$,
■ sodium sulphate, $Na_2SO_4$,
■ sodium thiosulphate, $Na_2S_2O_3$,
■ sodium peroxodisulphate, $Na_2S_2O_8$ (sodium persulphate).
In some cases you may not be able to interpret the reactions fully, but be sure to record your observations clearly and precisely. Ask your teacher which of the reactions you need to learn.

**Requirements**
- safety spectacles
- protective gloves
- 10 test-tubes in rack
- sodium sulphite solution, 0.2 M $Na_2SO_3$
- sodium sulphate solution, 0.2 M $Na_2SO_4$
- sodium thiosulphate solution, 0.2 M $Na_2S_2O_3$
- sodium peroxodisulphate solution, 0.2 M $Na_2S_2O_8$ (persulphate)
- hydrochloric acid, dilute, 2 M HCl
- Bunsen burner and bench protection mat
- wash-bottle of distilled water
- wood splints
- potassium dichromate(VI) solution, acidified, 0.1 M $K_2Cr_2O_7$
- strips of filter paper
- silver nitrate solution, 0.1 M $AgNO_3$
- 'silver residues' bottle
- iodine solution, 0.2 M $I_2$ in KI(aq)
- potassium iodide solution, 0.5 M KI
- iron(III) chloride solution, 0.5 M $FeCl_3$
- sodium hydroxide solution, 2 M NaOH
- test-tube holder
- spatula
- sodium sulphite-7-water, $Na_2SO_3 \cdot 7H_2O$
- sodium sulphate-10-water, $Na_2SO_4 \cdot 10H_2O$
- sodium thiosulphate-5-water, $Na_2S_2O_3 \cdot 5H_2O$
- sodium peroxodisulphate, $Na_2S_2O_8$
- boiling tube
- sulphur, powdered, S
- filter funnel and papers

## HAZARD WARNING

Aqueous sodium hydroxide is corrosive.

Silver nitrate is corrosive and toxic.

Sodium peroxodisulphate is a powerful oxidant.
Therefore you **must**:
- **wear safety spectacles and protective gloves;**
- **work at a fume cupboard (for test 1, which produces a toxic gas);**
- **smell gases cautiously (especially if you suffer from asthma).**

**Procedure** Perform tests 1 to 5 on separate portions (about 1 cm³) of **solutions** of the four oxosalts provided.

1. Working at a fume cupboard, add an equal volume of dilute hydrochloric acid, a few drops at a time. Warm **gently** (do not boil) and test any gas that you can see or smell. Record your observations in a copy of Results Table 5 and infer what you can from them.

2. Add about 2 cm³ of silver nitrate solution, a few drops at a time. Pour residues into the bottle provided.

3. Add a few drops of iodine solution.

4. Add a few drops of potassium iodide solution.

5. Add two drops of iron(III) chloride solution, followed by a few drops of dilute hydrochloric acid. Warm gently (do not boil) for half a minute and then add sodium hydroxide solution.

6. Heat small separate portions (about 0.5 g) of the **solid** oxosalts in a series of clean test-tubes. Test any gases evolved. Record your observations and inferences in your copy of Results Table 5.
7. Place about 0.5 g of finely powdered sulphur in a boiling tube and add about 10 cm$^3$ of sodium sulphite solution. Boil the mixture for 2–3 minutes and then filter.
8. Perform tests 1 to 5 on portions of the filtrate from step 7. Record your observations in the last column of Results Table 5 and attempt to identify the new species, X, in the filtrate.

**Results Table 5**

| Test | $Na_2SO_3$ | $Na_2SO_4$ | $Na_2S_2O_3$ | $Na_2S_2O_8$ | X |
|---|---|---|---|---|---|
| 1. Dilute hydrochloric acid. Warm | a | b | c | d | |
| 2. Silver nitrate solution | e | f | g | h | |
| 3. Iodine solution in aqueous potassium iodide | i | j | k | l | |
| 4. Potassium iodide solution | m | n | o | p | |
| 5. Iron(III) chloride solution. Acidify warm, add alkali | q | r | s | t | |
| 6. Effect of heat on solid | u | v | w | x | |

*Specimen results on page 113*

**Questions**

*Answers on page 114*

1. With the aid of your textbook(s) interpret your observations as far as you can, writing equations where possible, especially for (a), (c), (i), (k), (l) and (q). Insert oxidation numbers for sulphur in the equations.
2. Which of the reactions illustrate(s) disproportionation?
3. Which of the oxosalts is:
   i)   the strongest oxidising agent,
   ii)  the strongest reducing agent, and
   iii) the most stable salt?
4. Apart from the reactions used in the experiment, describe, from your previous knowledge, how you would distinguish between sodium sulphite and sodium sulphate.
5. What is the identity of the new species, X, produced by heating sodium sulphite solution with sulphur?

In the next exercise you investigate the structures of the sulphate, sulphite and thiosulphate ions.

**EXERCISE 64**

*Answers on page 116*

**a** Draw some 'electron-pair' structures for $SO_4^{2-}$ to explain why the ion is symmetrical in shape.
**b** What shape would you expect for the $SO_3^{2-}$ ion?
**c** What new bonds appear in the formation of thiosulphate ions by heating aqueous sulphite ions with sulphur?

Now you can apply your knowledge of the chemistry of sulphur to selenium.

**EXERCISE 65**

*Answers on page 116*

To answer this question, use the information provided below on selenium and its compounds.

Selenium occurs below sulphur in Group VI of the Periodic Table. It has an atomic number of 34 and a relative atomic mass of 79. The successive molar ionisation energies, in $10^3$ kJ mol$^{-1}$, are

0.94    2.1    3.1    4.1    7.1    7.9    15

Crystals of selenium dioxide contain chains of selenium and oxygen atoms:

$$-\underset{\underset{O}{|b}}{Se}\overset{a}{-}O\overset{a}{-}\underset{\underset{O}{|b}}{Se}\overset{a}{-}O\overset{a}{-}\underset{\underset{O}{|b}}{Se}\overset{a}{-}O\overset{a}{-}\underset{\underset{O}{|b}}{Se}$$

The bond lengths are $a = 0.178$ nm, $b = 0.173$ nm.
Selenium forms compounds analogous to sulphites and sulphates.

$$H_2SeO_3(aq) + 4H^+(aq) + 4e^- \rightleftharpoons Se(s) + 3H_2O(l) \qquad E^\ominus = +0.74 \text{ V}$$
$$SeO_4^{2-}(aq) + 4H^+(aq) + 2e^- \rightleftharpoons H_2SeO_3(aq) + H_2O(l) \qquad E^\ominus = +1.15 \text{ V}$$

**a** Explain the pattern in the first seven ionisation energies of selenium.
**b** i)  Draw a diagram to show the electron structure ('dot-and-cross') of selenium dioxide. Only the outer electronic shells of one unit of the chain need be shown.
   ii) What reason can you suggest for the bond lengths being different?

**c** Insert in (a copy of) the chart of redox potentials the information relating to the reduction of $H_2SeO_3$ to Se, and the reduction of $SeO_4^{2-}$ to $H_2SeO_3$.

**Figure 11**

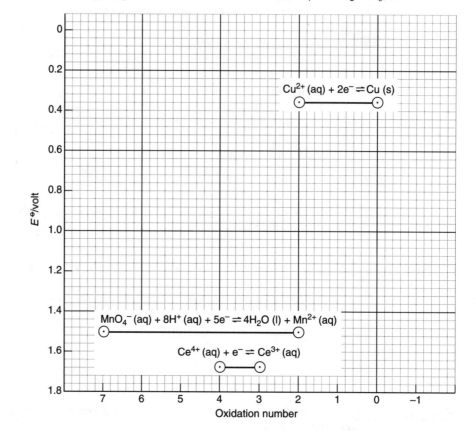

**d** Using the information displayed on the chart, answer the following questions:
  i)   What would you expect to **observe** if acidified $KMnO_4$ were added to $H_2SeO_3$ solution?
  ii)  Write a balanced equation for the reaction between $Ce^{4+}$ and $H_2SeO_3$ solutions.
**e** i)   Give the conventional cell diagram of a cell suitable for measurement of the standard e.m.f. of the reaction between Cu(s) and $SeO_4^{2-}$ (aq).
  ii)  Deduce the standard e.m.f. of such a cell.

Your teacher may suggest that you do the end-of-unit practical test (page 72) before you start on Part B. The test is based on the work you have done in Part A, and may be used for assessment purposes or as practice for students who will take a practical examination.

# ■ Part A test

Since your syllabus may not require you to study everything in Part A, we have decided to give you a selection of questions for each of the three Groups III, V and VI. Your teacher will tell you which questions to attempt, and the duration of the test.

You will not be required to attempt all the questions if you have studied all three groups.

**Section A – Group III**

1.  **a** Hydrated aluminium chloride is dissolved in water to form a colourless solution $X$. Treatment of $X$ with dilute ammonia solution gives a white precipitate, $Y$, which is insoluble in excess ammonia but does dissolve in dilute sodium hydroxide to form a colourless solution, $Z$.
    i)  Give the formulae of the principal aluminium-containing species in $X$, $Y$ and $Z$. (3)
    ii) What property of $Y$ is shown by its ability to dissolve in sodium hydroxide? (1)
    **b** Anhydrous aluminium chloride exists as covalently bonded molecules ($AlCl_3$ or $Al_2Cl_6$). Explain briefly why this molecule is not ionic. (3)
    **c** i)  Give and explain the **shape** of an $AlCl_3$ molecule.
    ii) Sketch the structure of an $Al_2Cl_6$ molecule to show the **bonding**. Why are the bond angles different from those in $AlCl_3$? (6)
    **d** (Omit this part if you have not done the relevant organic chemistry.) $AlCl_3$ is used as a catalyst in the reaction of benzene with chlorine to form chlorobenzene. The function of the $AlCl_3$ is to generate the electrophile $Cl^+$ thus:

$$AlCl_3 + Cl_2 \rightarrow Cl^+ + AlCl_4^-$$

    i)  What feature of the $AlCl_3$ molecule makes it suitable for this reaction?
    ii) By analogy with the reaction of benzene with the electrophile $NO_2^+$, deduce a mechanism for the chlorination of benzene. (4)

2.  In the vapour phase, the relative molecular masses of the chlorides of boron and aluminium are 117.3 and 267.0 respectively. (Relative atomic masses: $Al = 27.0$, $B = 10.8$, $Cl = 35.5$)
    **a** Give the molecular formulae of the two chlorides in the vapour phase. (2)
    **b** Using the usual symbols for covalent and coordinate bonds, draw a diagram to show the bonding in aluminium chloride vapour. (2)
    **c** Explain, by reference to aluminium oxide, the meaning of the term 'amphoteric oxide'. Give appropriate equations. (2)
    **d** i)  $Al^{3+}$ salts are common but $B^{3+}$ ions are unknown. Explain how this is consistent with the positions of the elements in Group 3 of the Periodic Table.
    ii) Explain why solutions of simple $Al^{3+}$ salts are acidic. (5)
    **e** Explain why aluminium saucepans should not be washed in strongly alkaline solutions. (2)

3.  In 1988, 20 tonnes of aluminium sulphate were accidentally tipped into drinking water near Camelford in Cornwall.
    **a** i)  Give the formula of the principal aluminium species present in an aqueous solution of aluminium sulphate.
    ii) Write an equation to show why the solution of aluminium sulphate is acidic.
    iii) Explain briefly why a solution of aluminium sulphate is more acidic than a solution of magnesium sulphate when the solutions are equimolar with respect to metal ions.
    iv) When magnesium turnings are added to aluminium sulphate solution, a gas is evolved and a white precipitate formed. Identify, by formula, the gas and the precipitate. (6)

**b** Describe what you would see, and give the formula of the final aluminium-containing product obtained, when a solution of aluminium sulphate is treated, dropwise until in excess, with aqueous sodium carbonate. (3)

**c** Water for drinking purposes must have a pH close to 7. Select from the following compounds the one which would be the best to add to a solution of aluminium sulphate in order to give a neutral solution of pH 7. Explain your choice.

$$Na_2CO_3 \qquad CaCO_3 \qquad NaOH \qquad Ca(OH)_2 \qquad\qquad (3)$$

**Section B – Group V**

4. The table below gives some data about four oxides of nitrogen.

| Formula | $\Delta H_f^{\ominus}$ /kJ mol$^{-1}$ | Oxidation number of nitrogen | Colour of gas |
|---------|---------|---------|---------|
| $N_2O$ | +82.0 | | Colourless |
| NO | +90.4 | +2 | Colourless |
| $NO_2$ | +33.2 | | Brown |
| $N_2O_4$ | +9.2 | | Pale yellow |

**a** i) Complete the blank oxidation numbers in the table.
  ii) Write, including the $\Delta H_f$ value quoted in the table, a balanced equation for the formation of nitrogen monoxide, NO. Use your equation to explain the presence of this gas in car exhaust fumes.
  iii) Suggest a reason why dinitrogen oxide, $N_2O$, relights a glowing splint. (7)

**b** Two of the oxides can co-exist in the equilibrium:

$$N_2O_4(g) \rightleftharpoons 2NO_2(g)$$

  i) Calculate the enthalpy change, $\Delta H$, for the forward reaction. (2)
  ii) Sealed tubes containing $NO_2$ and $N_2O_4$ are left as shown in the diagram below.

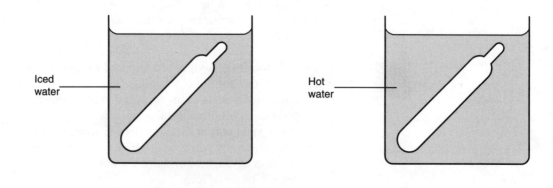

Iced water

Hot water

What colours would you expect to see in each tube?
Justify your answers with an explanation. (4)

5. This question concerns some chemistry of the elements in Group 5 of the Periodic Table.

   a Write down the electronic configurations of the nitrogen and the phosphorus atoms. (2)

   b Phosphorus forms a trichloride and a pentachloride, but nitrogen forms only a trichloride.
   
   i) Name the shapes of phosphorus trichloride and pentachloride molecules in the gas phase. (2)

   ii) Suggest **two** reasons why nitrogen does not form a pentachloride. (2)

   iii) Solid phosphorus pentachloride contains one octahedral and one tetrahedral ion. Suggest formulae for these ions. (2)

   c The enthalpy change for the decomposition of the trichlorides, $ACl_3$ (A = N or P), into their elements as represented by the equation:

   $$ACl_3(l) \rightarrow A_{(standard\ state)} + 1.5Cl_2(g)$$

   is $-258$ kJ mol$^{-1}$ for nitrogen and $+287$ kJ mol$^{-1}$ for phosphorus.

   i) Explain what is meant by the term **standard state**. (2)

   ii) Suggest **two** reasons for the difference in sign of the enthalpy changes of the reaction above for the two trichlorides. (2)

   iii) Explain what can be predicted about the stability of nitrogen trichloride at room temperature from the given enthalpy change. (2)

   iv) Why might the prediction in iii), above, not be borne out in practice? (1)

   d i) Describe, with the aid of a labelled diagram, how phosphorus trichloride may be prepared from phosphorus in the laboratory. (8)

   ii) Apart from the toxicity of chlorine, what other safety hazard is involved in this experiment? (2)

   iii) Briefly explain how phosphorus trichloride may be converted into phosphorus pentachloride. (2)

   e Write balanced equations for the complete reaction of the trichlorides of nitrogen and phosphorus with water. (3)

6. a i) Draw the molecular structure of white phosphorus. (2)

   ii) Write an equation for the reaction between white phosphorus and aqueous sodium hydroxide. (2)

   b i) Draw the molecular structure of the oxide which is formed when white phosphorus burns in an excess of oxygen. (2)

   ii) What is the oxidation state of phosphorus in this oxide? (1)

   iii) Write an equation for the reaction between the oxide and water. (2)

   c i) Predict the molecular shape of the chloride which is formed by the reaction between white phosphorus and an excess of chlorine. Explain your reasoning. (3)

   ii) How does this chloride react with water? (1)

   iii) Give an example of the use of this chloride in organic chemistry. (1)

   d Discuss briefly an environmental problem which arises from the large-scale use of phosphorus compounds. (2)

7.  **a** Using — to represent a covalent bond, draw the structures of nitrogen, oxygen and white (yellow) phosphorus molecules. (3)

    **b** Using the structures above and the following bond enthalpy values in kJ mol$^{-1}$, explain each of the following. (Calculations are not expected.)

    $$N_2(g) = 944 \qquad N—N = 163 \qquad O_2(g) = 496$$
    $$O—O = 146 \qquad P—P = 172 \qquad Cl—Cl = 242$$

    i)  Unsymmetrical dimethylhydrazine, $(CH_3)_2N—NH_2$, is so violently oxidised by dinitrogen tetroxide that the mixture is used as a rocket fuel, but nitrogen does not react at all with dinitrogen tetroxide.

    ii)  Oxygen oxidises iron(II) to iron(III), but only slowly, whereas hydrogen peroxide, $H_2O_2$, causes rapid oxidation.

    iii) White phosphorus ignites spontaneously in chlorine. (7)

    **c** The product of the reaction in (b) iii) reacts in the cold with more chlorine to give a pale yellow solid, Q.

    i)  Give the equation for this reaction.

    ii)  How does Q react with carboxylic acids? Explain how the product of this reaction is useful in synthetic organic chemistry. (4)

    **d** Fertilisers often contain phosphorus compounds, the proportion of phosphorus being expressed in terms of % $P_2O_5$ by mass.

    i)  Why would this **not** be the actual phosphorus compound in the fertiliser?

    ii)  If a fertiliser was listed as 10% $P_2O_5$ by mass but actually contained calcium phosphate, $Ca_3(PO_4)_2$, what would the composition be if expressed as % calcium phosphate by mass? (Relative atomic masses: O = 16, P = 31, Ca = 40) (6)

### Section C – Group VI

8.  **a** The recovery of sulphur from the hydrogen sulphide which is present in natural gas is now a major source of sulphur. Suggest a method by which this can be done. (6)

    **b** At temperatures below 369 K the most stable solid form of sulphur is rhombic sulphur, but at temperatures above that the most stable solid form is monoclinic sulphur.

    i)  What word is used to describe this phenomenon? (1)

    ii)  Mention two other elements which show the phenomenon and name the forms which are known. (3)

    iii) Suggest **two** ways in which monoclinic sulphur could be obtained from rhombic sulphur. (6)

    iv) What is the formula of the sulphur molecule in i) rhombic and ii) monoclinic sulphur? (1)

    These two forms differ in their densities. Why is this? (2)

    **c** When heated above 393 K sulphur melts to a free-flowing liquid. If the heating is continued, at about 453 K the viscosity suddenly increases by a factor of 2000, and the melt can then hardly be poured out of the containing vessel.

    Suggest a structural explanation for this strange behaviour. (5)

    **d** The mineral pyrite, $FeS_2$, often found in old coal-mine workings, is slowly converted by air, in the presence of water, into a solution of iron(II) sulphate in aqueous sulphuric acid,

    i)  Write a balanced equation for the reaction. (3)

    ii)  Why might the process cause environmental problems? (2)

    iii) What complex ion is present in the resultant solution? (1)

9. Sodium disulphate(IV) (metabisulphite), $Na_2S_2O_5$, is a white solid which is used in home-brewing to sterilise containers. When dissolved in water it reacts to form $HSO_3^-$ ions.

   **a** Deduce an equation for the reaction of $S_2O_5^{2-}$ ions with water. (1)

   **b** The aqueous solution containing $HSO_3^-$ ions evolves a pungent gas when acidified. Identify this gas and write an equation for its formation. (2)

   **c** On addition of aqueous, acidified potassium manganate(VII) the $HSO_3^-$ ions are oxidised to $SO_4^{2-}$ ions. Write half-equations for the oxidation and reduction processes and use these to construct an equation for the overall reaction. (3)

   **d** Potassium dichromate(VI) can also be used to oxidise $HSO_3^-$ ions to $SO_4^{2-}$ ions. The ionic equation for this reaction is:

$$Cr_2O_7^{2-} + 3HSO_3^- + 5H^+ \rightarrow 2Cr^{3+} + 3SO_4^{2-} + 4H_2O$$

   A tablet of sodium disulphate(IV) ($M_r = 190.0$), when dissolved in water, reacted with 17.55 $cm^3$ of 0.1000 M potassium dichromate(VI) solution. Use this information to determine the mass of sodium disulphate(IV) in the tablet. (4)

   **e** Give a reagent which can be used to distinguish between solid sodium disulphate(IV) and sodium sulphate. State your expected observations. (3)

10. When a solution containing a high concentration of sulphate ions is electrolysed under carefully controlled conditions, the ion $S_2O_8^{2-}$ is produced at the anode. Give a balanced half-equation for this reaction. (2)

   In this reaction either the sulphur atoms or the oxygen atoms might have been oxidised. Calculate the average oxidation numbers of sulphur and oxygen in the ion $S_2O_8^{2-}$ for both of these possibilities. (4)

   Suggest the displayed formula for the ion $S_2O_8^{2-}$, given that the ion contains one $O - O$ bond. On the basis of your displayed formula, explain the average oxidation numbers of sulphur and oxygen calculated previously. (4)

   The oxidising properties of the ion, $S_2O_8^{2-}$, were investigated. 25.0 $cm^3$ of a solution containing 0.0500 mol $dm^{-3}$ of $S_2O_8^{2-}$ was added to 10.0 $cm^3$ of a solution of potassium iodide, KI, of concentration 1.00 mol $dm^{-3}$. The iodine released required 20.0 $cm^3$ of a solution of sodium thiosulphate, $Na_2S_2O_3$, of concentration 0.125 mol $dm^{-3}$ for complete reaction. Calculate whether $S_2O_8^{2-}$ or iodide ions were present in excess and hence deduce the equation for the reaction of $S_2O_8^{2-}$ with iodide ions. (6)

   Would you expect $S_2O_8^{2-}$ ions to be able to oxidise $I_2$ to $IO_3^-$(aq) or to $H_5IO_6$? Justify your answer. (Use your data book.) (4)

# SOME INDUSTRIAL CHEMISTRY

In this part of the book we deal with some important industrial processes which involve p-block elements and their compounds.

For Group III we consider the extraction of aluminium from its ore; for Group V, the industrial manufacture of ammonia and nitric acid; and for Group VI, the manufacture of sulphuric acid. You will find it helpful to refer to ILPAC 8, Functional Groups, Chapter 2, where you considered some general points on industrial processes.

We begin with the extraction of aluminium.

CHAPTER

# THE EXTRACTION OF ALUMINIUM FROM ITS ORE

The importance of aluminium in our society has increased dramatically in the last hundred years. It has a unique combination of properties which make it useful in such diverse applications as overhead power cables, saucepans, aircraft bodies and milk-bottle tops. These and many other uses have made aluminium the next most important industrial metal after iron.

The rise in importance has been made possible by the relatively cheap production of the metal from its ore.

**OBJECTIVES**   When you have finished this chapter you should be able to:
■ state how pure aluminium oxide is obtained from **bauxite**;
■ explain how the **metal is extracted from its oxide by electrolysis**;
■ explain why this method of extraction is chosen;
■ explain how and why aluminium is **anodised**;
■ list the **uses of aluminium** and relate these to its **properties**.

The following passage is taken from *The Essential Chemistry Industry*, published by the Chemical Industry Education Centre at the University of York. It discusses the production of pure aluminium oxide from bauxite and the subsequent electrolysis of aluminium oxide to produce aluminium. However, the first part deals with the uses of aluminium to show why its production is so important.

Read the passage **carefully** and then do the exercises that follow. You are not expected to remember all the details of the process. The exercises are designed to focus your attention on the more important aspects.

# ALUMINIUM

## Uses

Because it can be readily fashioned or formed by almost every known metal-working process aluminium has a very wide range of applications.

It is malleable and easily worked. Also, it has a low density and, if properly alloyed, can be made stronger than structural steel. It is corrosion resistant and a good conductor of both heat and electricity. All the above properties can be improved or augmented (with the exception of corrosion resistance) by alloying aluminium with small amounts of other metals.

One of aluminium's strengths is that it is easily recycled at low cost (about 5% of the energy needed to produce primary metal from its ore is required) and approximately 60% of European consumption is recycled metal.

Thus aluminium is extensively used in building and construction work (wall facings, roofing, windows, doors), in transport (car wheels, bumpers, radiators, engines, container vehicle bodies, superstructures of ships, aeroplanes and railway rolling stock) and for the manufacture of consumer durables, containers and packaging (beverage cans, caps and foil).

In addition, aluminium is widely used in the electricity industry for long-distance transmission lines, meter casings and capacitor foil.

Aluminium chemicals are also widely used in industry. Aluminium hydroxide, $Al(OH)_3$, is extensively used as a flame retardant in polymers and elastomers. When subjected to heat aluminium hydroxide decomposes endothermically to alumina and steam thereby hindering the onset of fire. Aluminium hydroxide is also used in paper making as a high whiteness and brightness filler and coating additive.

Aluminium sulphate, $Al_2(SO_4)_3$, is prepared by reacting aluminium hydroxide with sulphuric acid. It is widely used in the water industry for removing colloidal impurities such as clays, humic substances and pathogenic bacteria from potable water. Aluminium sulphate is also used extensively in the purification of waste waters, including sewage, where it can precipitate phosphates thereby helping to prevent eutrophication of lakes, rivers and estuaries.

Activated alumina is prepared by heating aluminium hydroxide under carefully controlled conditions to impart porosity within the crystals and active sites on the surface. Activated aluminas are used as catalysts, catalyst supports and adsorbents in a wide range of industries including oil refining.

Alumina, aluminium oxide, $Al_2O_3$, is prepared from the hydroxide by calcination. The industrial applications of alumina centre on its physical properties of hardness and thermal and electrical insulation. Hence it is used widely in polishing compounds, abrasives, ceramics, refractories and electrical applications.

**Production (primary aluminium)**

| | |
|---|---|
| UK | 294 000 tonnes |
| World | $16.2 \times 10^6$ tonnes |

## Raw materials

The principal raw material is bauxite, a high alumina rock, containing 50-70% alumina. The alumina occurs as the trihydroxide ($Al(OH)_3$ (gibbsite), and as the monohydroxide, $AlO(OH)$ (boehmite and diaspore). The other necessary aluminium compound is cryolite $Na_3AlF_6$. Limited deposits of cryolite occur only in Greenland but these are no longer mined commercially. All cryolite now used is produced synthetically.

The extraction process consumes substantial quantities of sodium hydroxide, fuel both for heating in the preliminary bauxite treatment and for calcination of the hydroxide to produce alumina. Carbon, derived from petroleum coke and coal tar pitch, is used for anodes in the reduction process which consumes approximately 400 kg carbon per ton of metal produced.

## Cryolite synthesis

Cryolite may be synthesised in a variety of ways. One of these is to react HF with sodium aluminate solution and to precipitate cryolite by adding sodium carbonate:

$$NaAlO_2 + 6HF + Na_2CO_3 \rightarrow Na_3AlF_6 + 3H_2O + CO_2 \text{ (at 323–343 K)}$$

Aluminium fluoride is used to adjust the electrolyte composition and to make up fluoride losses during electrolysis (see below). Aluminium fluoride is produced by the reaction:

$$Al_2O_3 + 6HF \rightarrow 2AlF_3 + 3H_2O \text{ (at 675–875 K)}$$

The aluminium industry is one of the largest consumers of hydrogen fluoride.

## The process

Aluminium smelting is a very energy intensive process requiring 14–15 kWh of electricity to produce 1 kg of aluminium. Cheap electrical power is therefore economically essential and consequently aluminium smelters are located where hydroelectric power is available or near coal or natural gas sources.

There are two main stages in the extraction:

### Bauxite purification

The principal impurities in bauxite are iron(III) oxide (3–25%), silica (1–7%) and titanium(IV) oxide (2–3%). To remove them, powdered bauxite is mixed with approximately 10% sodium hydroxide solution and the resulting mixture heated under pressure (4 atm) at about 420 K. The digestion takes about 1–2 hours, the exact conditions employed depending on the nature of the original bauxite.

$$Al(OH)_3 + NaOH \rightleftharpoons NaAl(OH)_4$$

The aluminium hydroxide content of the bauxite dissolves but the oxides of iron and titanium remain insoluble. Some of the silica can dissolve which is undesirable and process conditions are chosen to minimise this event. The insoluble residue (red mud) is settled, washed to recover caustic soda, filtered and dumped, usually to landfill.

The sodium aluminate is clarified and pumped to large precipitator tanks which may be over 24 m high and have capacities of 1000 $m^3$ or more. The solution is cooled and seeded with aluminium hydroxide crystals and stirred for up to three days allowing crystallisation to occur in a controlled manner so as to achieve the optimum particle size distribution for processing in subsequent stages.

$$NaAl(OH)_4 \rightleftharpoons Al(OH)_3 + NaOH$$

The aluminium hydroxide is separated from the sodium hydroxide by filtration. A proportion of the aluminium hydroxide is retained as seed and the sodium hydroxide is recycled. The remaining aluminium hydroxide is calcined in rotary kilns at about 1270 K to produce alumina.

$$2\, Al(OH)_3 \rightarrow Al_2O_3 + 3H_2O$$

## Electrolytic reduction of aluminium oxide to aluminium (The Hall–Héroult process)

The commercial process for reducing alumina to aluminium is an electrolytic process invented independently, and simultaneously, by Paul Héroult in France and Charles Hall in America in 1886.

A reduction cell (see Fig. 12, page 57) is a shallow open-topped steel box approximately 8 m by 4 m by 1 m deep, the floor of which is lined with graphite carbon which acts as the cathode connection. In operation the box is filled with molten cryolite (at 1225 K) which acts as the electrolyte. The side walls of the box are lined with carbon and designed so that cryolite freezes in a thin layer on their surface. Suspended in the electrolyte are several blocks of graphitic carbon which form the anode. The key to the process is the use of cryolite which dissolves alumina and thus enables it to be electrolysed.

The exact nature of ions in the electrolyte is not fully known but it is thought that alumina dissolves to form an oxy-fluoride ion, $Al_2OF_6^{2-}$, which is discharged at the anode as follows:

$$2Al_2OF_6^{2-} + C \rightarrow 4AlF_3 + CO_2 + 4e^- \qquad (1)$$

The overall reaction at the cathode is probably:

$$AlF_6^{3-} + 3e^- \rightarrow Al + 6F^- \qquad (2)$$

Molten aluminium is more dense than cryolite and collects on the cell bottom where it forms the operating cathode. It is normal to operate a cell with a pool of aluminium 10 cm deep with metal being syphoned off daily to maintain this level. The metal is at least 99% pure with small amounts of iron and silicon being the main impurities.

The process operates continuously with fresh alumina being added regularly to maintain its concentration in the electrolyte at 2–5 wt% [between 2% and 5% by weight].

Carbon is consumed in the anode process and so the anode blocks must be renewed regularly. Typically a cell has twenty anode blocks one being replaced each day in a twenty day cycle (to replace all of the blocks at once would result in too much cooling of the cell), with the block size being designed to give an operating life of 20–21 days.

Modern cells operate at 150 000–300 000 amps and 4.0–4.5 volts, producing 1–2 tonnes of aluminium per day. About one third of the electrical energy is used in the reduction process, the remainder being used to maintain the cell at its operating temperature. A typical factory will contain nearly 400 such cells and produce 200 000 tons of metal per year. Associated with it will be a factory to fabricate anodes (about 100 000 tons of anodes will be required) and a casthouse to produce alloyed metal in a form which can be used by other factories to produce sheet, foil, extrusions etc.

Metal of up to 99.999% purity is produced by further electrolytic refining and/or zone refining. This is principally used by the electronics industry for capacitor foil, hard disc drives and conductor tracks on silicon chips.

## Aluminium and the environment

### Bauxite mining

Almost all bauxite mines are of the open-cut type where shallow deposits can be removed without underground mining. Most mines replace vegetation, prevent soil erosion and restore wildlife habitat. Ultimate land use dictates whether native species are used for reclamation or whether the land is upgraded through restoration.

### Bauxite residues (red mud)

After processing the bauxite in the alumina plant, the residue material is washed before disposal to recover the sodium hydroxide. Because complete removal is not possible, the residues are typically managed in dyked disposal areas which can range from ten to several hundred hectares in size. After the residue impoundment is full, the deposit can be reclaimed and re-vegetated.

### Aluminium smelters

Fluoride emissions during electrolysis can amount to 20–30 kg per ton of aluminium produced. The use of sophisticated fume capture equipment, associated with fume dry scrubbing equipment, dramatically reduces fluoride emissions into the environment down to a level of under 1 kg of total fluoride per ton of aluminium produced. With sound environmental management involving proper control facilities and technology, the modern aluminium smelter can be operated with minimum environmental problems.

The first exercise on the passage above concerns the uses of aluminium.

**EXERCISE 66**
*Answers on page 117*

**a** What properties of aluminium make it particularly useful for each of the following applications?
  i) Long-distance overhead electricity cables.
  ii) Drinks cans.
  iii) Windows and doors.
  iv) Aircraft construction.
**b** How does aluminium hydroxide work as a flame retardant? Write an equation.
**c** At low concentrations, aluminium compounds in water appear to be non-toxic (aluminium has been used for many years in cooking utensils and food containers) but higher concentrations cause problems. These problems became very evident when 20 tonnes of aluminium sulphate, $Al_2(SO_4)_3$, were accidentally tipped into the wrong tank at a waterworks in Cornwall in 1988, contaminating the local water supply. For what two purposes might the aluminium sulphate have been intended?

The next exercise focuses on the extraction process. You will need to refer to Fig. 12.

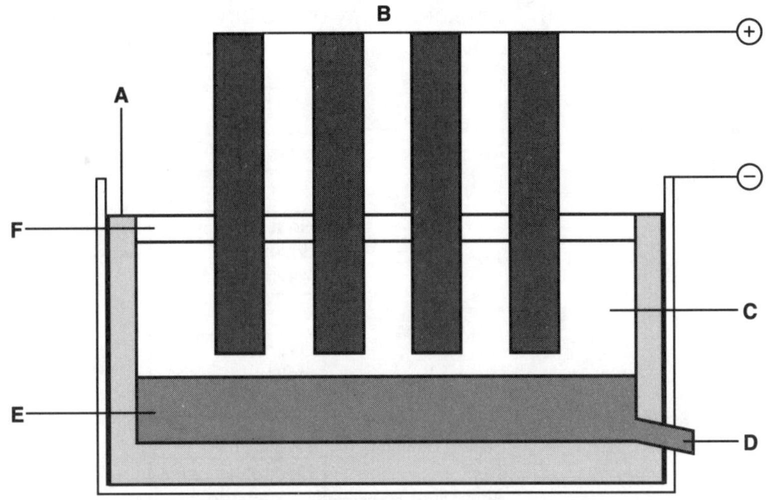

**Figure 12**
The Hall–Héroult cell for the production of aluminium.

**EXERCISE 67**

*Answers on page 117*

a  In addition to bauxite, name three substances essential to the process. Which one of these is not actually consumed in great quantity?

b  Summarise briefly how the main impurities are removed from bauxite. Rewrite in ionic form the relevant equation given in the text.

c  What is the main constituent of 'red mud', that gives it its colour?

d  What are the three essential operations that produce pure aluminium oxide from the solution of sodium aluminate?

e  Complete the labelling on a copy of Fig. 12.

f  Many textbooks state that the reactions during electrolysis are:

at the cathode $Al^{3+} + 3e^- \rightarrow Al$
at the anode $O^{2-} \rightarrow \frac{1}{2}O_2 + 2e^-$
followed by $C + O_2 \rightarrow CO_2$

The overall process is then summarised as:

$$2Al_2O_3 + 3C \rightarrow 4Al + 3CO_2$$

The passage suggests that the electrode processes are certainly more complex but still not completely understood. However, you can show that the equations given in the passage lead to the same overall equation, as follows:

i)  Combine equations *(1)* and *(2)* for the electrode reactions into one.

ii)  Combine the result from i) with the equation:

$$AlF_3 + 3F^- \rightarrow AlF_6^{3-}$$

so that $AlF_3$ does not appear.

iii)  Combine the result from ii) with the equation for the reaction of alumina with cryolite as it dissolves:

$$Al_2O_3 + 4AlF_6^{3-} \rightarrow 3Al_2OF_6^{2-} + 6F^-$$

g  Why do you think aluminium is not produced by the electrolysis of pure molten aluminium oxide?

h  Alumina cannot be reduced by carbon directly in a process similar to the production of iron in a blast furnace. Look up values of $\Delta H_f^\ominus$ and use them to explain why this is so.

i  How is the electrolysis made into a continuous process?

**j** Complete a copy of the flow diagram in Fig. 13 which summarises the essential steps in the extraction of aluminium from bauxite.

**Figure 13**
Summary of aluminium production.

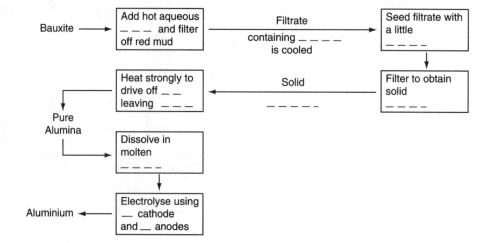

The next exercise is concerned mainly with the economic and social aspects of aluminium production.

**EXERCISE 68**

*Answers on page 119*

**a** What is the most important factor in the siting of an aluminium smelter?

**b** The UK (and many other industrial countries) has no substantial deposits of bauxite. What factors do you think would influence the decision whether to import bauxite or purified alumina for a proposed new smelter?

**c** Much more common than bauxite are a variety of aluminium containing clays, but it is at present not economic to extract aluminium from them. Nevertheless, much research work has been done on this alternative source of aluminium, particularly in Russia. Suggest reasons for this.

**d** Summarise the three main environmental problems associated with aluminium production.

The Alcan aluminium works in Fort William, Scotland.

You learned in Part A that the reactivity of aluminium is masked by a thin coherent layer of aluminium oxide. Without this protective layer, aluminium would not be such a useful metal. The range of uses can be extended still further by making the oxide layer much thicker in a process called anodising. Anodised aluminium is not only more resistant to corrosion, but it can be made more attractive by surface dyeing.

 In the next experiment you can anodise and dye a piece of aluminium. If you do not do this experiment we suggest that you read the procedure so that you know, in outline, how the process is carried out.

## EXPERIMENT 4   Anodising aluminium

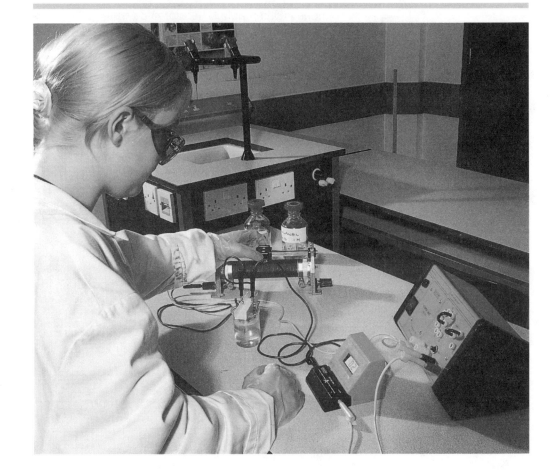

**Aim**   The purpose of this experiment is to demonstrate the effect of anodising aluminium on its electrical conductivity and behaviour towards dyes.

**Introduction**   Aluminium is anodised by making it the anode during the electrolysis of a solution which normally releases oxygen – you will use dilute sulphuric acid. Instead of oxygen, aluminium ions are released and these are immediately hydrolysed to give a layer of hydrated oxide. After electrolysis you compare anodised and unanodised aluminium by testing their conductivities and immersing them in a dye solution.

**Requirements**
- safety spectacles
- 2 pieces of sheet aluminium, 7 cm $\times$ 3 cm, mounted on wooden bars
- 2 beakers, 100 cm$^3$
- forceps or tweezers
- cotton wool
- propanone, $CH_3COCH_3$
- sulphuric acid, dilute, 1 M $H_2SO_4$
- Bunsen burner, tripod, gauze and bench mat
- thermometer, 0–100 °C
- d.c. supply, 12 V
- ammeter, 1 A
- rheostat, 10 $\Omega$, 4 A
- 4 connecting leads, two with crocodile clips at one end
- 2 large pins
- dye solution, e.g. alizarin red

**HAZARD WARNING**

Propanone is flammable. Therefore you **must**:
■ **keep the bottle away from flames and return its stopper as soon as possible.**

Dilute sulphuric acid is corrosive. Therefore you **must**:
■ **wear safety spectacles.**

**Procedure
– Part A**

**Anodising**

1. Mark one of the wooden supports for the electrodes with a 'plus' sign. This carries the piece of aluminium to be used as the anode.
2. Working at a fume cupboard, remove any grease from the surface of the anode by swabbing with cotton wool soaked in propanone and held in forceps or tweezers. From now on, hold the anode only by its wooden support.
3. Pour about 75 cm$^3$ of dilute sulphuric acid into a small beaker, heat to about 40 °C and stand it on the bench.
4. Place the electrodes in the beaker and set up the circuit shown in Fig. 14. Make sure the clean anode is connected to the positive terminal of the supply.
5. Switch on the d.c. supply and adjust the rheostat to give a current of 0.3 A. Maintain this current for at least 15 minutes (longer if it is convenient), adjusting the rheostat if necessary to keep the current constant.
6. Switch off the supply. Disconnect the electrodes, remove them from the beaker and wash them with distilled water. Has their appearance changed?

**Figure 14**

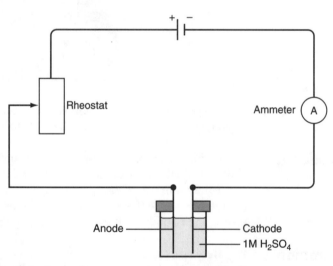

**– Part B**

**Conductivity testing**

7. Clip a large pin to each of the leads you had previously connected to the electrodes. Adjust the rheostat so that when the two pins touch each other, the current is no greater than 1 A.
8. Touch each pin **lightly** on the surface of the cathode and note the ammeter reading. Repeat for the anode. Is there any difference? Now press the pins more firmly on to the anode surface. If this does not change the ammeter reading, try scratching the surface with the pins.

**– Part C**

**Dyeing**

9. Place spots of dye solution on both anode and cathode and leave for 2–3 minutes. (If you have enough dye, you can immerse the electrodes.) Rinse the electrodes with water and inspect them.

**Questions**

*Answers on page 119*

1. How does anodising affect aluminium with regard to its:
   **a** appearance,
   **b** conductivity,
   **c** susceptibility to dyes?
2. You probably noticed that you had to reduce the variable resistance during the electrolysis in order to prevent the current from falling. Suggest two reasons for this.
3. In what applications might anodising be:
   **a** an advantage,
   **b** a disadvantage?

To consolidate your knowledge of aluminium production and its uses we suggest you attempt the following teacher-marked exercise. You will need to read a little more about the uses of aluminium, looking particularly for the relationships between uses and properties.

**EXERCISE**

*Teacher-marked*

**a** How would you anodise a piece of aluminium? Discuss the principles underlying this form of corrosion protection of the metal.
**b** Give an account of the uses of aluminium, emphasising its advantages and disadvantages in each case.

The next exercise illustrates another aspect of the extraction of aluminium. If your syllabus includes the interpretation of Ellingham diagrams, you will have studied them in ILPAC 11, Transition Elements, in connection with the extraction of iron. Revise this work, if necessary, or omit the exercise if it does not relate to your syllabus.

**EXERCISE 69**

*Answers on page 119*

**Figure 15**

Study the Ellingham diagrams in Fig. 15 and answer the questions that follow.

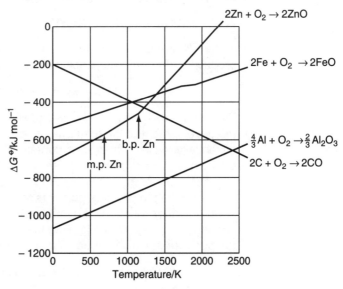

**a** State the temperatures above which carbon will reduce the following oxides, and explain your answers:
   i)   ZnO,
   ii)  FeO,
   iii) $Al_2O_3$.
**b** Explain why aluminium and iron are extracted by different methods.

You have now completed the sections on the industrial chemistry of the Group III element, aluminium. In the following chapters we turn to Group V and consider first the industrial production of ammonia.

# THE INDUSTRIAL PRODUCTION OF AMMONIA

You probably know something about this topic from your earlier studies, in which case some of the work you do now will be revision.

**OBJECTIVES**

When you have finished this chapter you should be able to:
- describe the **Haber process** for the production of ammonia from nitrogen and hydrogen;
- state the **sources** from which **nitrogen and hydrogen** are obtained;
- explain the **choice of temperature and pressure** in the process;
- state the **effect of the catalyst** on the equilibrium position and on the rate at which equilibrium is reached;
- list some of the important **uses of ammonia**.

Read about the Haber process for the manufacture of ammonia, paying particular attention to the choice of temperature and pressure. You may need to refer to your notes on ILPAC 6, Equilibrium I: Principles, to remind yourself about Le Chatelier's principle before attempting the next two exercises.

Exercise 70 is based on the following extract from an article, 'The Fixation of Nitrogen', in an education journal, *Education in Chemistry*, by S P S Andrew, July 1978. Read the passage **carefully**. Although it was written some time ago, the principles are still largely relevant.

The fixation of atmospheric nitrogen in the form of ammonia is one of the foundations of the modern chemical industry. Combined in the form of ammonium nitrate, urea or ammonium phosphate, over 50 million tonnes* per year are used worldwide as fertilisers. Other uses of fixed nitrogen amount to some 10 million tonnes* per year, explosives, dyestuffs and polymer manufacture predominating. [*These figures had more than doubled by 1994.]

## Ammonia production

In the UK ammonia currently is entirely made using natural gas (methane) to produce hydrogen. The simplified overall stoichiometry of ammonia production is:

$$(\tfrac{1}{2}N_2 + \tfrac{1}{8}O_2) + \tfrac{5}{8}H_2O + \tfrac{7}{16}CH_4 \rightarrow NH_3 + \tfrac{7}{16}CO_2 \qquad \Delta H^\ominus = -6.7 \text{ kJ mol}^{-1}$$

The use of methane to make hydrogen for ammonia production was the result of the development of the methane-steam reforming process. This process essentially makes a hydrogen-containing gas by reacting methane with steam over a nickel-on-**refractory** catalyst. The methane required by this stoichiometry is sufficient in theory to supply the energy for the complete chemical transformation to produce gaseous ammonia. In practice extra fuel, amounting to about a further 50 per cent over the stoichiometric, must be used for raising the temperature of the gases leaving the steam reformer to about 1100 °C, a temperature at which the equilibrium of the reaction:

$$CH_4 + H_2O \rightleftharpoons CO + 3H_2 \text{ (The methane-steam reaction.)}$$

is displaced well to the right.

## Synthesising ammonia

Two factors are of crucial importance in synthesis – firstly, the equilibrium of the reaction, secondly, its rate. The reaction equilibrium:

$$\tfrac{1}{2}N_2(g) + \tfrac{3}{2}H_2(g) \rightleftharpoons NH_3(g) \qquad \Delta H^\ominus = -47.3 \text{ kJ mol}^{-1}$$

is markedly temperature and pressure sensitive [Fig. 16]. The rate of the reaction is determined by the efficacy of the synthesis catalyst.

Since the early days of commercial ammonia synthesis this catalyst has been metallic iron promoted by potassium hydroxide (KOH) and containing a small amount of mixed refractory oxides such as alumina ($Al_2O_3$), silica ($SiO_2$) and magnesium oxide (MgO). Fabrication of the catalyst is by melting together the ingredients in their oxide forms and casting the molten iron oxide ($Fe_3O_4$) melt. The resulting solid sheet is then broken up to give 5–10 mm chunks. Finally, it is reduced. In its active state the catalyst consists predominantly of iron crystallites of a few hundred Ångstrom (1 Å = 0.1 nm) units in size. These crystallites are separated by amorphous refractory oxides and partially covered by alkali promoter.

In order to maintain catalyst activity it is necessary to eliminate catalyst poisons from the gas entering the ammonia synthesis reactor. Hydrogen sulphide and water vapour chemisorb [that is bond strongly] on the surface of the iron rendering it ineffective as a catalyst. In addition, water results in a recrystallisation of the catalyst structure giving larger iron crystallites and hence an even lower exposed iron surface area with consequent loss in catalytic activity. Water vapour can enter the catalyst either as itself or as carbon monoxide (CO) or carbon dioxide ($CO_2$), for both of these gases will be hydrogenated to methane and water vapour over the catalyst. Rigorous exclusion of hydrogen sulphide ($H_2S$), water, carbon monoxide (CO) and carbon dioxide ($CO_2$) from the catalyst is, therefore, desirable.

The commercial operating range of ammonia synthesis catalyst is between 400 °C and 540 °C. Below 400 °C, the catalyst is not sufficiently active and above 540 °C it loses surface area by sintering [partial melting causing particles to fuse together] too rapidly. In order, therefore, to obtain reasonable conversions per pass through the converter, synthesis pressures in modern plants fall in the range 8080–35 350 kPa (80–350 atm).

Contrary to what might at first be thought, this choice of synthesis pressure is little affected by considerations of the total mechanical power requirements for compressing the nitrogen and hydrogen feed as, over a wide range of pressures (10 100–30 300 kPa: 100–300 atm), the sum of the power required for compression plus power for driving a refrigeration system for condensing the product liquid ammonia plus power for returning unconverted hydrogen and nitrogen to the converter remains virtually constant. The lower gas compression power of low pressure synthesis is almost exactly balanced by the higher refrigeration power.

The use of lower synthesis pressures has been greatly influenced by the availability of suitable cheap rotary gas compressors. Steam turbine drives are usually employed for these compressors, the steam being almost wholly generated by heat rejected from the process. By thus utilising waste heat for driving the machinery of the plant, the only thermal inefficiency is the result of having to burn methane in the reformer furnace. The energy efficiency of the whole process of making ammonia is, therefore, about 65 per cent.

**Figure 16**

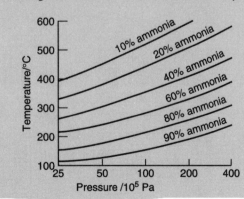

**EXERCISE 70**

*Answers on page 119*

Answer the following questions, which relate to the preceding extract.

**a i)** Give the names and formulae of **two** nitrogen-containing compounds used as fertilisers.

**ii)** Give the name of a widely used nitrogen-containing polymer.

**b i)** Outline the steam-reforming of methane.

**ii)** Explain the meaning of the word **refractory**.

**iii)** Deduce the sign of the enthalpy change for the following reaction with reasoning:

$$CH_4(g) + H_2O(g) \rightarrow CO(g) + 3H_2(g)$$

**iv)** For the equilibrium:

$$CH_4(g) + H_2O(g) \rightleftharpoons CO(g) + 3H_2(g)$$

give an expression for the equilibrium constant, $K_p$, and explain what happens to the equilibrium partial pressure of the hydrogen when the total pressure of the steam is increased.

**c i)** What **two** factors are of crucial importance in the synthesis of ammonia?

**ii)** Describe the nature and fabrication of the catalyst used in ammonia synthesis.

**iii)** What effect does water vapour have on the catalyst? How does it enter the catalyst?

**d** Using the information given in the passage, discuss how the economy of the ammonia synthesis depends upon the catalyst, power requirements, pressure and temperature.

In the next exercise also you need to refer to the extract preceding Exercise 70.

**EXERCISE 71**

*Answers on page 121*

Carbon monoxide is removed from the synthesis gas mainly by means of the 'shift reaction', in which it reacts with steam over a hot catalyst:

$$CO(g) + H_2O(g) \rightleftharpoons CO_2(g) + H_2(g)$$

**a** Why is it important to remove carbon monoxide from the synthesis gas?

**b** What is another useful result of employing the shift reaction?

**c** $CO_2$ must also be removed but this is much easier. How would you suggest it could be done?

The next exercise deals with the conditions of temperature and pressure employed in the Haber Process.

Ammonia production (ICI).

**EXERCISE 72**

*Answers on page 121*

In the Haber process for the manufacture of ammonia, the reactants enter the reaction vessel in the ratio one mole of nitrogen to three moles of hydrogen. The equation is:

$$N_2(g) + 3H_2(g) \rightleftharpoons 2NH_3(g) \qquad \Delta H^\ominus = -100 \text{ kJ mol}^{-1} \text{ at 200 atm and 800 K}$$

After leaving the reaction vessel, the gases pass over a heat exchanger before the ammonia is removed and unreacted gases recycled. Argon has to be 'blown off' from time to time.

The percentages of ammonia present in the equilibrium mixture under different conditions, are shown in the graph.

**Figure 17**

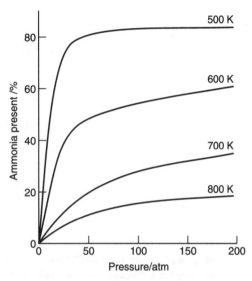

In practice, the process is operated at 200 atm and 800 K.
**a** Give **one** reason for:
  i) **not** using a higher temperature than 800 K,
  ii) **not** using a lower temperature than 800 K.
**b** Give **one** reason for **not** using a pressure exceeding 200 atm.
**c** How could ammonia be separated from the other two gases on the output side of the plant?
**d** The heat exchange ensures that the heat of the reaction is not wasted, but does useful work elsewhere. Give one other reason why the heat exchanger is important to the efficiency of the reaction process.
**e** i) What is the equilibrium constant expression in terms of $p_{N_2}$, $p_{H_2}$, $p_{NH_3}$?
  ii) How would the presence of argon affect the partial pressure of ammonia and hence the yield?
  iii) Where does the argon come from?
**f** The equilibrium partial pressure of nitrogen is 42 atm out of a total pressure of 200 atm.
  i) What is the partial pressure of hydrogen?
  ii) Calculate the partial pressure of ammonia and the percentage of ammonia in the equilibrium mixture.
  iii) Give **one** reason why the commercial process is more economical when the reaction is **not** allowed to reach equilibrium.

Eighty percent of the ammonia produced by the Haber process is used in fertilisers: some liquid ammonia is injected directly into the soil, some is converted to carbamide (urea), $CO(NH_2)_2$, but most is converted to ammonium salts such as ammonium sulphate and, particularly, ammonium nitrate. For the latter, nitric acid is required and this is also manufactured from ammonia.

# THE MANUFACTURE OF NITRIC ACID

Much of the ammonia produced by the Haber Process is used to manufacture nitric acid, which is used as an oxidising agent and nitrating agent as well as an acid. It is required for making both inorganic and organic nitrates, such as ammonium nitrate and ethyl nitrate, and organic nitro-derivatives such as trinitrotoluene (TNT) and nitroglycerine. These products include useful fertilisers and explosives.

**Figure 18**
Uses of nitric acid.

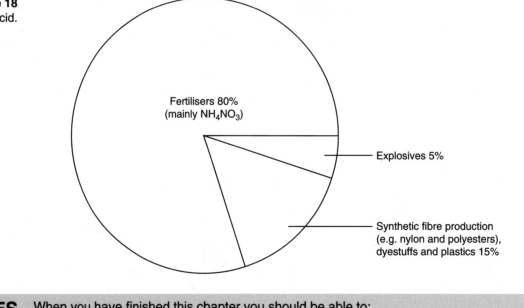

Fertilisers 80%
(mainly $NH_4NO_3$)

Explosives 5%

Synthetic fibre production
(e.g. nylon and polyesters),
dyestuffs and plastics 15%

## OBJECTIVES

When you have finished this chapter you should be able to:
■ describe the **manufacture of nitric acid**;
■ use **Le Chatelier's principle** to explain the operating conditions;
■ state the main **uses of nitric acid**.

In the next exercise, the process is described for you but you apply your knowledge of physical chemistry to answer the questions.

**EXERCISE 73**
*Answers on page 122*

Nitric acid is manufactured by the catalytic oxidation of ammonia and the subsequent reaction of the product with water. The ammonia gas is mixed with a large excess of air and passed at atmospheric pressure through a grid made up of layers of gauze of composition platinum 90%, rhodium 10%. The reaction takes place at about 900 °C. Some recent manufacturing plants, which can be rather smaller, operate this stage at a pressure of 8 atmospheres.

$$4NH_3(g) + 5O_2(g) \rightleftharpoons 4NO(g) + 6H_2O(g) \qquad \Delta H_{1200}^{\ominus} = -903 \text{ kJ mol}^{-1}$$

The gases leaving the converter are cooled and mixed with more air to promote a second oxidation reaction.

$$2NO(g) + O_2(g) \rightleftharpoons N_2O_4(g) \qquad \Delta H_{400K}^{\ominus} = -116 \text{ kJ mol}^{-1}$$
$$N_2O_4(g) \rightleftharpoons 2NO_2(g) \qquad \Delta G_{400K}^{\ominus} = -13 \text{ kJ mol}^{-1}$$

The reaction mixture is then compressed to 8 atmospheres and passed up absorption towers 30 metres high, down which water is flowing:

$$3NO_2(g) + H_2O(l) \rightleftharpoons 2HNO_3(aq) + NO(aq) \qquad \Delta H_{298K}^{\ominus} = -135 \text{ kJ mol}^{-1}$$

The nitric acid flowing out of the absorption tower is coloured green due to dissolved nitrogen oxide.

**Figure 19**

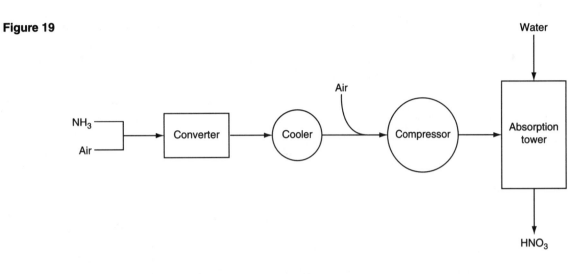

**a** Write down the equilibrium constant expression for $K_p$ for the reaction between ammonia and oxygen.

**b** Give **two** reasons why **excess** air is used in the plant to oxidise the ammonia gas.

**c** Why do you think the first stage is operated at **atmospheric** pressure?

**d** Why do you think the first stage is operated at a temperature of 900 °C?

**e** Why do you think the gases are cooled before the second oxidation?

**f** How will the ratio of dinitrogen tetraoxide to nitrogen dioxide be affected by a rise in temperature? Give a reason. (If you have not studied $\Delta G$, assume that $\Delta H$ values are similar.)

$$N_2O_4(g) \rightleftharpoons 2NO_2(g) \qquad \Delta G^{\ominus}_{450K} = -22 \text{ kJ mol}^{-1}$$

**g** State **two** significant factors that a manufacturer should have considered before building a plant to operate at 8 atmospheres throughout.

**EXERCISE 74**
*Answers on page 122*

Nitric acid is manufactured from the air together with a suitable source of hydrogen; ammonia is an essential intermediate.

**a** Write the equations, including state symbols, for **four** chemical reactions which are important in this manufacture of nitric acid. For each reaction give the conditions under which it is carried out.

**b** A sample of gas taken from one stage of the process had the composition, by volume: $H_2$, 74.2%; $N_2$, 24.7%; $CH_4$, 0.8%; Ar, 0.3%; CO, less than one part per million. Do you think this would be a suitable general composition for the gas entering the ammonia synthesiser? Explain.

We now turn our attention from Group V to Group VI, where the most important industrial chemistry concerns the production of sulphuric acid.

# THE MANUFACTURE OF SULPHURIC ACID BY THE CONTACT PROCESS

Sulphuric acid is a very important industrial chemical. Many processes require it at some stage of the manufacture. It is used to make fertilisers such as ammonium sulphate and calcium superphosphate, and in the making of explosives, detergents, accumulators, varnishes and artificial fibres such as rayon.

**Figure 20**
The uses of sulphuric acid.

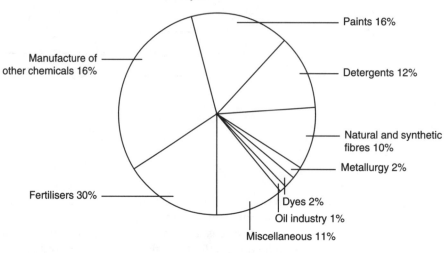

**OBJECTIVES**  When you have finished this chapter you should be able to:
- outline the **Contact process** for the production of **sulphuric acid** (full technical details are not required);
- state the main sources from which **sulphur dioxide** is obtained;
- explain the **choice of temperature and pressure** in the contact process;
- state the **effect of the catalyst** on the equilibrium;
- list the main **uses of sulphuric acid**.

You probably already know something about the contact process from your earlier studies, in which case some of the work that follows will be revision. Refer to your notes and/or your textbooks to help you with the next three exercises, the first of which refers to the flow chart in Fig. 21 below.

**Figure 21**
The production of sulphuric acid.

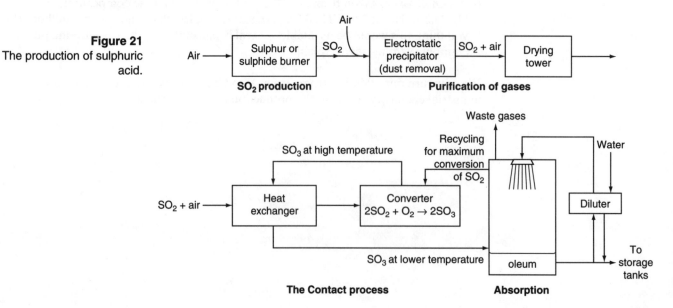

**EXERCISE 75**
*Answers on page 123*

**a** State briefly how sulphur dioxide is obtained from:
i)   sulphur,
ii)  a sulphide ore,
iii) impurities in supplies of crude oil and natural gas.
Give the appropriate equations in your answer.
**b** There have been periods in the past when a significant proportion of the UK's sulphuric acid was manufactured from sulphur extracted from anhydrite, $CaSO_4$, a widely occurring and easily obtained mineral. Suggest reasons why this process was introduced and then discarded.
**c** Name the catalyst used in the Contact process.
**d** Why is it important to remove impurities from the sulphur dioxide?
**e** Explain why sulphur trioxide is absorbed into concentrated sulphuric acid and then diluted with water rather than absorbed into water immediately. Give any relevant equation.
**f** Why is the sulphur trioxide passed through a heat exchanger?
**g** In older plant employing the Contact process, 98% conversion of sulphur dioxide was achieved in a series of catalyst beds in the converter without recycling the mixed gases. Nowadays 99.5% conversion is achieved by absorbing sulphur trioxide before the final catalyst bed. The cost of doing this is not covered by the small extra production of sulphuric acid, so why do you think it is done?

In the next exercise you consider the effects of temperature, pressure and a catalyst on the conversion of sulphur dioxide to sulphur trioxide.

**EXERCISE 76**
*Answers on page 123*

The following equilibrium is involved in the preparation of sulphuric acid by the Contact process:

$$2SO_2(g) + O_2(g) \rightleftharpoons 2SO_3(g) \qquad \Delta H^{\ominus} = -188 \text{ kJ mol}^{-1}$$

**a** Write the expression for $K_p$ for the above equilibrium.
**b** State the effect on the equilibrium of each of the following changes:
i)   increase of pressure at constant temperature,
ii)  increase of temperature at constant pressure.
Explain your answer in each case.
**c** Give approximate values of the temperature and pressure that are used in the industrial process.
**d** Comment on the pressure used in the industrial process.
**e** A catalyst containing vanadium(V) oxide is used in the process. What is its effect on the conversion of $SO_2$ to $SO_3$?

Now try another A-level question on the Contact process.

**EXERCISE 77**
*Answers on page 124*

**a** Sulphuric acid is manufactured from sulphur by the 'contact' process. Write the equations, including state symbols, for **four** chemical reactions which are important in this process of manufacture. For each reaction give the conditions under which it is carried out.
**b** On one occasion a sample of the gas mixture in one part of the plant was analysed and found to consist of $SO_2$, 10%; $SO_3$, 0.2%; $O_2$, 11%; and $N_2$, 79% by volume.
i)   At what point in the process was the sample taken?
ii)  Does the general composition of the sample indicate that the plant was working properly or not? Explain.

Environmental problems have been mentioned briefly in preceding chapters. In the next chapter we pay them a little more attention.

# ENVIRONMENTAL PROBLEMS

We have already dealt with the pollution of air by sulphur dioxide and oxides of nitrogen in ILPAC 5, Introduction to Organic Chemistry (Chapter 6). There we showed that the main sources of these gases in the air were fossil fuels burnt in power stations, and the internal combustion engine. However, you will have realised that some emissions occur in the manufacture of sulphuric acid and nitric acid, although legislation has led to design modifications that have greatly reduced these emissions. One environmental concern we have not dealt with elsewhere is the use of nitrogenous fertilisers, which has increased enormously in recent years. This increased use has brought great benefits in increased food production to supply expanding populations, but not without some accompanying problems.

**OBJECTIVE**    When you have finished this chapter you should be able to:
■ summarise the main **environmental problems** associated with the manufacture of ammonia, nitric acid, sulphuric acid and fertilisers.

Read about the consequences of using large quantities of nitrogenous fertiliser. Look for a brief account of the possible dangers of **high levels of nitrate** in rivers and other water supplies, both to human health and to the environment generally. For most syllabuses you do not need much detail – if in doubt, ask your teacher. Then attempt the following exercises.

**EXERCISE 78**
*Answers on page 124*

a Why does river water in agricultural areas sometimes contain high concentrations of nitrate ions?
b What does the term 'eutrophication' mean, and how is it linked to nitrate ions?
c Agricultural fertilisers also contain phosphates, which are also linked with eutrophication. Suggest two reasons why phosphates are less of a problem than nitrates in this connection (but see Exercise 41 for a different situation).
d Much of our drinking water comes from rivers but another important source is the underground water stored in suitable rock strata called aquifers. Aquifers also become contaminated with nitrate ions, though much more slowly. Why is the presence of nitrate ions in drinking water thought to be a possible health hazard?

**EXERCISE 79**
*Answers on page 124*

Sulphur dioxide is a major pollutant in the atmosphere. The quantity of sulphur dioxide in a sample of air can be determined by reaction with hydrogen peroxide and titrating the sulphuric acid formed with standard alkali.

a Write the equation for the reaction between sulphur dioxide and hydrogen peroxide.
b In a typical experiment 200 $m^3$ of polluted air was bubbled slowly through 25.0 $cm^3$ of a solution of hydrogen peroxide. The resulting solution was neutralised by 15.3 $cm^3$ of sodium hydroxide solution of concentration 0.100 mol $dm^{-3}$.
   Calculate the sulphur dioxide concentration, in g $m^{-3}$, in the sample of air. ($A_r(S) = 32.0, A_r(O) = 16.0$)

Before you tackle the end-of-unit test, your teacher may want you to do one or more of the following teacher-marked exercises.

**EXERCISE**
*Teacher-marked*

1. Many important industrial processes involve equilibrium reactions. Write an account of chemical equilibrium, placing it as far as possible in an industrial context. You should certainly consider the meaning and usefulness of an equilibrium constant, the role of catalysts, and how the application of Le Chatelier's principle allows a qualitative interpretation of the effect of temperature and pressure on the equilibrium concerned.

**EXERCISE**

*Teacher-marked*

2. **a** Ammonium nitrate is used as a fertiliser. Give an account of its manufacture **from its elements**. For all three industrial steps confine your answer to essential conditions and equations.

   **b** Give one element (other than N) essential to plant growth and a compound containing it which is used as a fertiliser.

   **c** Mention some environmental problems which may be associated with the large-scale use of fertilisers.

**EXERCISE**

*Teacher-marked*

3. **a** Give an account of the conditions under which **either** ammonia **or** sulphuric acid are manufactured on a large scale. Include the starting materials, the conditions under which the reaction takes place, the nature of the catalyst, and the approximate yield.

   **b** Explain, by considering the effect on the position of equilibrium and the cost of operating the plant, why the temperature and pressure used are as you have described them.

   **c** Outline **two** major uses of **either** ammonia **or** sulphuric acid.

**EXERCISE**

*Teacher-marked*

4. **a** Describe:
    i) the synthesis of ammonia from nitrogen and hydrogen,
    ii) the oxidation of ammonia to nitric acid.

   **b** i) Give an account of pollution arising from nitrogen-containing compounds.
    ii) Outline methods of controlling such pollution.

## ■ End-of-unit tests

To find out how well you have learned the material in this book, particularly in Part B, try the test on page 74. Ask your teacher which questions you should attempt.

There is also an end-of-unit practical test below, which is more relevant to Part A. This test can be used as practice for a practical examination, or for an assessment of your skills at recording and interpreting observations. Therefore we do not include in this book any specimen results or a requirements list. There are no special hazard warnings but you should wear safety spectacles for all practical work. This test should be completed in 90 minutes.

## ■ End-of-unit practical test

You are provided with aqueous solutions, labelled B, C and D, of three salts. They are compounds of the same three elements.

Carry out the following experiments. Safety spectacles should be worn throughout. Record your observations and inferences in (a copy of) the table opposite, commenting on the types of chemical reactions involved.

**Results Table 6**

| Test | Observations | Inferences |
|------|--------------|------------|
| 1. To 2 or 3 cm$^3$ of solution B add a few drops of dilute hydrochloric acid. Now add a few drops of aqueous barium chloride | | |
| 2. To 2 or 3 cm$^3$ of solution B add a few drops of aqueous lead ethanoate (lead acetate). Now add an excess of ammonium ethanoate (ammonium acetate) and shake the mixture | | |
| 3. To about 5 cm$^3$ of solution C in a boiling tube add about 10 cm$^3$ of dilute hydrochloric acid. Allow the mixture to stand for a minute or two. Cautiously smell the mixture. Warm, if necessary, and test the gas evolved. Describe below how you performed this test<br><br>**Method** | | |
| 4. **a** In a small beaker place 2 or 3 cm$^3$ of aqueous copper(II) sulphate. Acidify with two drops of dilute sulphuric acid. Now add about 5 cm of aqueous potassium iodide<br>**b** To the mixture obtained in 4(a) add solution C until in excess, swirling well | | |
| 5. **a** In a small beaker mix about 2 cm$^3$ of aqueous sodium chloride with an equal volume of aqueous silver nitrate<br>**b** To the mixture obtained in 5(a) add solution C until in excess, swirling well | | |
| 6. Dissolve about 1 g of iron(II) sulphate crystals in dilute sulphuric acid and add a few drops of aqueous potassium thiocyanate. Now add a few drops of solution D | | |
| 7. **a** In a small beaker place about 5 cm$^3$ of aqueous potassium iodide and a few drops of dilute sulphuric acid. Now add about 5 cm$^3$ of solution D. Warm this mixture<br>**b** To the mixture obtained in 7(a) add solution C until in excess, swirling well | | |
| 8. Evaporate a small quantity of solution B to dryness and perform a flame test on the residue. Describe below how you do this<br><br>**Method** | | |

# ■ End-of-unit test

To find out how well you have learned the material in this book, try the test that follows. Read the notes below before starting.

1. Do not attempt *all* the questions. Ask your teacher which ones are appropriate.
2. Hand your answers to your teacher for marking.

1. **a** Aluminium and its alloys have replaced iron and steel in many applications. Suggest **one** advantage and **one** disadvantage of using aluminium compared to steel in the manufacturing industry. (2)
   **b** Give an account of the following points concerning the chemistries of aluminium and iron:
   i)   the charges on, and colours of, their common cations; (3)
   ii)  the reactions of aluminium oxide with aqueous sodium hydroxide and with hydrochloric acid; (3)
   iii) the reactions between aqueous potassium thiocyanate and aqueous solutions of iron salts. (2)
   In your answer, including equations for all the reactions you describe and relate the above properties of aluminium and iron to the electronic structures and sizes of their ions.

2. Aluminium sulphate and calcium oxide are sometimes added to water supplies to co-precipitate suspended solids and bacteria. A small amount of aluminium remains in solution and its presence in drinking water may contribute to the mental illness known as Alzheimer's disease.
   **a** Suggest an identity for the insoluble aluminium compound that might be formed when water, aluminium sulphate and calcium oxide are mixed, and write a balanced equation for its production. (2)
   **b** Why is it important not to add too much calcium oxide? (2)
   **c** By considering the uses to which aluminium metal is put, suggest another possible source of ingestion of aluminium. (1)

3. **a** Nitrogen can be obtained in the laboratory by warming a mixture of ammonium chloride and sodium nitrite, $NaNO_2$. Steam is also produced and a solid is left.
   i)   Suggest an identity for the solid and write a balanced equation for the reaction. (2)
   ii)  Calculate the changes in oxidation numbers of the nitrogen atoms during this reaction. (2)
   iii) A similar reaction takes place when ammonium chloride is heated with sodium nitrate, $NaNO_3$, but this time the only different product is an oxide of nitrogen.
        Suggest a formula for this oxide, and calculate the oxidation number of nitrogen in it. (2)
   **b** One of the main uses of nitrogen compounds is for agricultural fertilisers.
   i)   Name **two** compounds used in this way. (1)
   ii)  Why is it sometimes necessary to apply nitrogen-based fertilisers several times during the growing season, whereas one application of a phosphate-based fertiliser is all that is likely to be needed? (1)
   iii) What are the environmental consequences of the over-use of nitrogen-based fertilisers? How do they arise? (2)

4. This question is concerned with the purity of water supplies. Nitrate is one of the major impurities found in water supplies.
   **a** i) Give **one** possible source of nitrate in a water supply. (1)
      ii) Give **one** possible consequence of the presence of nitrate in a water supply.(1)
   **b** The recommended maximum concentration of nitrate ($NO_3^-$, molar mass = 62 g mol$^{-1}$) in tap water is 50 ppm (i.e. parts per million) or 0.050 g dm$^{-3}$. What is the nitrate concentration of such a solution expressed in mol dm$^{-3}$? (1)
   **c** The nitrate ions can be removed and replaced by chloride ions by passing the water through a suitable ion exchange material containing chloride ions. For each nitrate ion absorbed one chloride ion is released.
       If 1.0 kg of the ion exchange material contains 1.0 mol of exchangeable chloride ions, what is the minimum mass of the material (assuming 100% efficiency of exchange) needed to remove the nitrate dissolved in 1000 dm$^3$ of the water obtained from a borehole that contains 50 ppm of nitrate? (3)
   **d** Although this process can produce water that is essentially free of nitrate, the water is usually mixed with untreated water so that the tap water supplied to the consumer contains about 40 ppm of nitrate. Suggest **one** reason why this is done. (1)

5. **a** Most of the earth's nitrogen exists uncombined as nitrogen gas in the atmosphere. What is the reason for this? (2)
   **b** The reaction used in the industrial production of ammonia from atmospheric nitrogen is represented by the equation:

   $$N_2 + 3H_2 \rightleftharpoons NH_3 \qquad \Delta H_f^{\ominus} = -92 \text{ kJ mol}^{-1}$$

     i) What are the effects of changes in temperature and pressure on the **equilibrium** yield of this reaction? Explain your answer. (4)
     ii) What sort of conditions are employed in industry to achieve the best **economic** yield from the process? Explain your answer. (4)
   **c** i) State the major use of ammonia and its derivatives. Why is this so important to the world economy? What environmental problem is caused by this use? (4)
     ii) Give one other important use of ammonia in the chemical industry. (1)
   **d** Suggest **two** reasons why liquid ammonia is a good solvent for many inorganic compounds and also for some organic compounds such as phenols and carboxylic acids. (2)
   **e** Draw the structure of the species which would be formed by reaction between ammonia and $Cu^{2+}$(aq). (3)

6. The industrial synthesis of ammonia is carried out in two main stages. In the first overall stage, hydrogen is produced from methane and water in 'reforming reactions'. Hydrogen is then catalytically combined with nitrogen from the air in the second stage.
   In a typical primary steam reforming reaction, methane and steam are passed over a nickel catalyst in a furnace maintained at 850 °C and 30 atm pressure, resulting in the following reaction.

   $$CH_4(g) + H_2O(g) \rightleftharpoons CO(g) + 3H_2(g) \qquad \Delta H^{\ominus} = +210 \text{ kJ mol}^{-1} \qquad (1)$$

   Air is then injected into the gas mixture from the primary reformer. The oxygen from the air reacts with some hydrogen in the mixture according to the following reaction.

   $$2H_2(g) + O_2(g) \rightarrow 2H_2O(g) \qquad \Delta H^{\ominus} = -482 \text{ kJ mol}^{-1} \qquad (2)$$

Unconverted methane in the mixture from the primary reformer reacts with steam from reaction *(2)* in a secondary reformer by the process in equation *(1)*.

**a** The methane feedstock often contains sulphur compounds. Why is it necessary to remove these before treatment of the feedstock in the reformers? (1)

**b** What is the principal economic advantage of running reactions *(1)* and *(2)* in the same plant? (2)

**c** Why is there some unconverted methane in the gases from the primary reformer? (1)

**d** i) Write an equation for the Shift reaction, which is used to remove carbon monoxide from the gases produced in the reformers.

　 ii) What economic advantage has the Shift reaction other than in removing carbon monoxide?

　 iii) Give a disadvantage of the Shift reaction. State, giving the chemical equation for the reaction involved, the treatment needed to overcome this disadvantage. (5)

**e** Final traces of carbon dioxide and carbon monoxide are removed by catalytic reduction by hydrogen in the gas mixture.

　 i) Give the equation for the removal of carbon dioxide.

　 ii) Give the equation for the removal of carbon monoxide. What is the value of the standard enthalpy change $\Delta H^{\ominus}$ for this process? (3)

**f** Suggest **one** impurity which is still present in small quantities in the final mixture of hydrogen and nitrogen. (1)

7. **a** Write a balanced equation for the synthesis of ammonia from its elements. (2)

　 **b** i) Name the catalyst used in industry to speed up this process. (1)

　　 ii) State the other reaction conditions used in industry for this process. (2)

　 **c** Deduce the oxidation state of nitrogen in:

　　 i) hydrazine, $N_2H_4$, 　　　　　 ii) nitric acid, $HNO_3$. (2)

　 **d** Ammonia dissolves in water to form an alkaline solution.

　　 i) Write a balanced equation to illustrate this. (1)

　　 ii) Write a balanced equation for the neutralisation of aqueous ammonia by nitric acid. (1)

　　 iii) Calculate the volume of $0.20$ mol $dm^{-3}$ nitric acid required to react completely with 20 cm³ of $1.0$ mol $dm^{-3}$ aqueous ammonia. (2)

　　 iv) Suggest a use for the product of this reaction and comment on a possible disadvantage of this use. (3)

8. Ammonia is produced from its elements on a large scale using the Haber process.

　 **a** Write an equation for the formation of ammonia from its elements. (1)

　 **b** The formation of ammonia is an exothermic reaction. In choosing the conditions under which the reaction is to be performed, decisions as to pressure and temperature must be made on economic grounds. State the arguments which influence such decisions.

　　 i) Argument in favour of using a high pressure.

　　 ii) Argument against using a high pressure.

　　 iii) Argument in favour of using a high temperature.

　　 iv) Argument against using a high temperature. (4)

　 **c** For reasons of environmental safety the concentration of ammonia in the air downwind of an ammonia production plant was measured by the following procedure.

　　 A 20 000 litre (measured at s.t.p.) sample of the air was slowly bubbled through an excess of dilute hydrochloric acid. The resulting solution was made alkaline and heated, the ammonia liberated being dissolved in exactly 50 cm³ of 0.1 M hydrochloric acid, which is a large excess. 40.00 cm³ of 0.1 M sodium hydroxide solution were required to neutralise the excess of acid.

　　 Calculate the concentration of ammonia in the air in units of moles of ammonia per litre of air. (5)

9. In this country, nitrogen is usually fixed by the Haber process. In some other countries (for example those with hydro-electric power) use is made of a process whereby massive electric sparks are passed through air, causing nitrogen and oxygen to combine:

$$N_2(g) + O_2(g) \rightleftharpoons 2NO(g) \qquad \Delta H^\ominus = +180 \text{ kJ mol}^{-1} \qquad \text{Reaction 1}$$

   a  What does **fixation** of nitrogen mean? (1)
   b  Predict the effect on the equilibrium yield of NO in **reaction 1** of increasing the temperature (as in the electric spark). (3)
   c  The nitrogen monoxide reacts with air at lower temperatures to form nitrogen dioxide. This dissolves in water to form nitric acid, releasing more nitrogen monoxide. Write equations for these reactions. (2)
   d  Why is the process of **reaction 1** only used in countries with hydro-electric power? (1)

10. a  The initial stage in the manufacture of nitric acid is the catalytic oxidation of ammonia by air at 850 °C using a platinum–rhodium alloy gauze as catalyst. The reaction is:

$$4NH_3(g) + 5O_2(g) \rightleftharpoons 4NO(g) + 6H_2O(g) \qquad \Delta H = -905 \text{ kJ mol}^{-1}$$

   State why:
   i)   the gauze is heated to start the reaction,
   ii)  the gauze remains hot during the reaction,
   iii) platinum and rhodium are used although they are expensive metals. (3)
   b  In the second stage of the manufacture of nitric acid, the gases from the catalyst unit are cooled to 150 °C when nitrogen dioxide is formed.

$$2NO(g) + O_2(g) \rightleftharpoons 2NO_2(g) \qquad \Delta H = -113 \text{ kJ mol}^{-1}$$

   Explain why the gases are cooled. (2)
   c  The nitrogen dioxide in (b) is then absorbed in water. The initial reaction is:

$$2NO_2(g) + H_2O(l) \rightarrow HNO_3(aq) + HNO_2(aq)$$

   Name this type of reaction and give reasons for your answer. (3)
   d  Give **two** major uses of nitric acid (2)
   e  State **one** fixed cost and **one** variable cost incurred in nitric acid production. (2)

11. The conditions predicted by theory for maximum conversion of reactants into products are not always used by industrial manufacturers. The exothermic reaction:

$$2SO_2(g) + O_2(g) \rightleftharpoons 2SO_3(g)$$

   is carried out at a pressure just above atmospheric using a series of catalyst beds which operate at temperatures between approximately 675 K and 725 K.
   a  Explain how theoretical considerations indicate that a high pressure should be used for this reaction.
   b  Give **one** reason why a high temperature is used.
   c  The reaction mixture is cooled between catalyst beds. Give **two** reasons for this.
   d  $SO_3$ is removed from the gas stream by absorption in 98% $H_2SO_4$. Explain why acid, rather than water, is used.
   e  Most modern plants use a double absorption process. State what is meant by **double absorption**, and give **two** reasons why it is used. (10)

12. This question concerns the industrial production of sulphuric acid.

$$\text{Stage 1} \quad S + O_2 \rightarrow SO_2$$
$$\text{Stage 2} \quad 2SO_2 + O_2 \rightleftharpoons 2SO_3$$
$$\text{Stage 3} \quad SO_3 + H_2O \rightarrow H_2SO_4$$

  **a** i) For each of the three stages in this process describe briefly the conditions
      under which that stage is carried out. (6)
    ii) For stages 2 and 3 outline the reasons for choosing those conditions. (5)
  **b** Why is it that this industrial process is economically viable even though the
    product, sulphuric acid, is sold relatively cheaply? (2)
  **c** A particular chemical company operates a sulphuric acid plant located in a small
    town in the centre of England. The company would not build the plant in the
    same place today.
    i) Suggest **two** possible reasons why the sulphuric acid plant should be located in
      such an unlikely place. (2)
    ii) Suggest **two** possible reasons why the company might choose **not** to expand
      that facility despite an increased demand for the product. (2)
  **d** Give **three** important uses of sulphuric acid which show how it contributes to
    improving our standard of living. (3)

13. **a** Sulphuric acid is an important feedstock (raw material) for the chemical industry.
    Its preparation from sulphur involves converting the sulphur into sulphur
    dioxide, and then oxidising this gas to sulphur trioxide. The sulphur trioxide is
    dissolved in concentrated sulphuric acid and then mixed with water.
      Outline the method used industrially in converting sulphur into each of the two
    oxides mentioned above. You should give reaction conditions where possible. (5)
  **b** Elemental sulphur reacts readily with metals to form sulphides. A sulphide is
    precipitated when hydrogen sulphide gas is bubbled through a solution of a
    suitable metal salt.
    i) Write equations to illustrate these two reactions.
    ii) The sulphide ion is a strong base. Use this fact to help explain why a solution
      of sodium sulphide in water is noticeably alkaline.
    iii) Hydrogen sulphide is much less soluble in hot water than in cold water. Hence
      deduce what happens if a solution of sodium sulphide in water is boiled for an
      extended period. (8)

14. **a** Explain the reasons for the physico-chemical conditions used in the Contact
    Process for the manufacture of sulphuric acid from sulphur dioxide and oxygen.
    What environmental problems arise from the Contact Process? (10)
  **b** Describe and explain how you would show that sulphuric acid is dibasic. (2)
  **c** Briefly outline a method of preparation of each of the following compounds:
    i) hydrated $CuSO_4(s)$ from $CuO$,
    ii) $PbSO_4(s)$ from $PbO$.
    Confine your answers to stating essential conditions and how to obtain a pure
    product safely. (8)

15. **a** The oxidation of sulphur dioxide to sulphur trioxide is an important industrial
    process:

$$2SO_2(g) + O_2(g) \rightleftharpoons 2SO_3(g) \qquad \Delta H^{\ominus}(298\ K) = -196\ kJ\ mol^{-1}$$

    i) What operating conditions are used in practice to achieve the best **economic**
      yield from the process? Give reasons. (4)
    ii) Explain how the sulphur trioxide is converted to sulphuric acid. (2)

iii) Iron pyrites (iron disulphide, $FeS_2$) is oxidised by roasting in air to form iron(III) oxide, used to make iron, and sulphur dioxide which is used to make sulphuric acid. If 80% of the sulphur in iron pyrites is ultimately converted to pure sulphuric acid, calculate the mass of $H_2SO_4$ produced from 1.00 tonne of pyrites. ($A_r(H) = 1$, $A_r(O) = 16$, $A_r(S) = 32$, $A_r(Fe) = 56$) (4)

iv) State **two** uses of sulphuric acid. (2)

**b** The amount of sulphur dioxide in the atmosphere is a matter of concern.

   i) Give **one** source of the sulphur dioxide entering the atmosphere and suggest how the chemical properties of sulphur dioxide might be used to limit its emission. (3)

   ii) State **one** of its adverse effects. (1)

16. Read the following account and answer the questions which follow.

**Fertilisers**

For healthy growth, crops require three major nutrient elements. These are nitrogen (N), phosphorus (P) and potassium (K). Synthetic fertilisers must provide these elements in a form which is readily taken up by crops.

A number of important intermediates such as ammonia, sulphuric acid, phosphoric(V) acid and nitric acid are required for fertiliser manufacture. A large fertiliser factory will therefore consist of a number of chemical plants to produce these intermediates as well as plants to make the fertilisers themselves. Such a factory can then produce both 'straight' N fertilisers (e.g. ammonium nitrate or carbamide (**urea**)), and 'compound' NPK fertilisers (e.g. a mixture of ammonium nitrate, triammonium phosphate and potassium chloride). Below is shown a flow diagram for an integrated fertiliser factory.

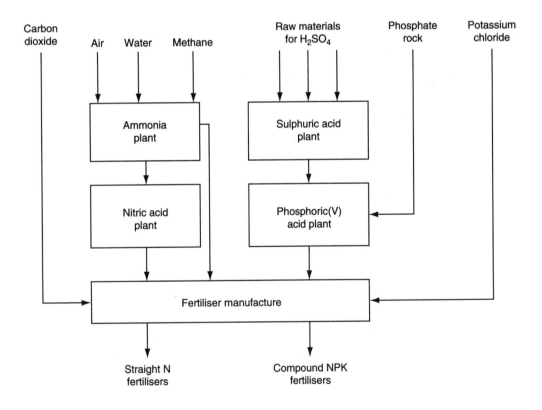

**a** Name **three** major raw materials needed for the manufacture of sulphuric acid. (2)

**b** It is proposed to produce the organic fertiliser carbamide (**urea**), $NH_2CONH_2$, from carbon dioxide and ammonia according to the reaction:

$$2NH_3(g) + CO_2(g) \rightleftharpoons NH_2CONH_2(g) + H_2O(g) \qquad \Delta H \text{ is negative}$$

i)   By considering equilibrium and kinetic factors, discuss how the optimum conditions of temperature and pressure might be chosen.

ii)  Calculate the minimum masses of ammonia and carbon dioxide required to produce 1 tonne of carbamide.

iii) Suggest a suitable source of carbon dioxide for the process.          (11)

**c** i)   Give **four** factors that would be important in deciding where to site a new fertiliser factory such as the one shown above.

ii)  Give **three** advantages of designing an integrated factory (such as the one shown above) rather than having a number of plants on separate sites.

iii) Most chemical plants are run continuously throughout the year. What implications might this have for a fertiliser manufacturer?          (9)

# ANSWERS

(Answers to questions from examination papers are provided by ILPAC and not by the examination boards.)

**EXERCISE 1**   **a** The ore is called bauxite and consists primarily of aluminium oxide (hydrated).

   **b** Boron is usually encountered as a brown amorphous powder (occasionally as black crystals) with very low conductivity. Aluminium is a silvery grey crystalline metal with high conductivity.

   **c** The differences suggest that boron is non-metallic and aluminium is metallic in character.

   **d** Assuming that the usual group trend of increasing metallic character continues down the group, we should expect thallium to be metallic in appearance and a good electrical conductor.

**EXERCISE 2**   **a** B: $1s^2 2s^2 2p^1$ or $(He)2s^2 2p^1$
       Al: $1s^2 2s^2 2p^6 3s^2 3p^1$ or $(Ne)3s^2 3p^1$

   **b** Boron and aluminium exhibit the +3 oxidation state.

   **c** By analogy with Group IV we might expect the 'inert pair effect' to be observed in Group III also, in which case the possible oxidation states of thallium would be +1 $(Tl^+)$ and +3 $(Tl^{3+})$.

   **d** i) $B^{3+}$: 0.016 nm  $Al^{3+}$: 0.045 nm
         These are among the smallest ionic radii listed.

      ii) In view of its high charge and very small size, we might expect $B^{3+}$ to polarise any neighbouring anions to such an extent as to make all boron compounds predominantly covalent. The same argument applies to aluminium, but to a lesser extent.

**EXERCISE 3**   In comparing two half-reactions, the more negative $E^\ominus$ value favours the oxidation process, and the more positive $E^\ominus$ value favours the reduction process. This is equivalent to the 'anti-clockwise rule', provided the half-equations are written in the same order as they appear in the redox series, i.e. with the more negative $E^\ominus$ value first.

**a**

$$Al^{3+}(aq) + 3e^- \rightleftharpoons Al(s) \qquad E^\ominus = -1.66 \text{ V}$$
$$2H^+(aq) + 2e^- \rightleftharpoons H_2(g) \qquad E^\ominus = 0.00 \text{ V}$$

$\Delta E^\ominus = (0.00 - -1.66) \text{ V} = \mathbf{+1.66 \text{ V}}$

Aluminium would be expected to react with dilute acids, producing aqueous aluminium ions and hydrogen gas.

**b**

$$Al(OH)_4^-(aq) + 3e^- \rightleftharpoons Al(s) + 4OH^-(aq) \qquad E^\ominus = -2.35 \text{ V}$$
$$2H_2O(l) + 2e^- \rightleftharpoons H_2(g) + 2OH^-(aq) \qquad E^\ominus = -0.83 \text{ V}$$

$\Delta E^\ominus = \mathbf{+1.52 \text{ V}}$

Aluminium would be expected to react with dilute alkalis, producing aqueous aluminate ions and hydrogen gas.

# EXPERIMENT 1

*Specimen results*

**Results Table 1**
Reaction of aluminium foil

| Reagent | Observations | Identity of any gas given off |
|---|---|---|
| Sodium hydroxide solution | Slow evolution of gas at first, becoming quicker | Hydrogen |
| Sodium hydroxide solution after immersion in $CuCl_2$(aq) | Rapid evolution of gas | Hydrogen |
| Dilute hydrochloric acid | No reaction in cold. Gas evolved on heating | Hydrogen |
| Dilute hydrochloric acid after immersion in $CuCl_2$(aq) | Gas evolved in the cold | Hydrogen |
| Air after immersion in $CuCl_2$(aq) | No visible reaction | – |
| Air after immersion in $HgCl_2$(aq) | White crumbly solid appeared on surface | – |

**Results Table 2**
Reactions of the solutions

| Reagent | Observations using solutions of | |
|---|---|---|
| | **Al in NaOH** | **Al in HCl** |
| Dilute sulphuric acid | A white gelatinous precipitate appeared which dissolved on shaking. More acid made the ppt. permanent and still more redissolved it. The tube became warm | |
| Sodium hydroxide solution | | A white gelatinous precipitate appeared which dissolved on shaking. More alkali made the ppt. permanent and still more redissolved it. The tube became warm |
| Sodium carbonate solution | | A gas was evolved ($CO_2$) and a white gelatinous precipitate appeared which dissolved on shaking. More alkali made the ppt. permanent but it did not redissolve. The tube became warm |

*Questions*

1. Aluminium is normally covered in a thin invisible layer of aluminium oxide which is not easily removed and protects the metal surface from further attack by air and even by dilute acids. Any oxide formed on the surface of iron does not adhere well enough to protect the metal. (Remember also that $E^{\ominus}$ values refer only to standard conditions.)

2. Solutions of copper(II) chloride and mercury(II) chloride remove the oxide layer and expose the aluminium surface, which then reacts more vigorously with acids and alkalis. (How the oxide layer is removed is not clear. It is interesting to note that the chloride ions play a part – other copper salts are not effective.)

3. Displacement reactions occur with the cations in solution, producing metallic copper and mercury. The copper is simply washed off and the aluminium surface then becomes again covered with a protective layer of oxide. On the other hand, liquid mercury dissolves some aluminium to form an alloy (called an amalgam) which exposes aluminium atoms to attack by oxygen but does not allow a coherent protective layer to form.

4. Washing soda and many oven-cleaners are alkaline in solution and therefore tend to attack aluminium pans and other kitchenware.

5. **a** For a transfer of six electrons:
$$2Al(s) \rightarrow 2Al^{3+}(aq) + 6e^-$$
$$6H^+(aq) + 6e^- \rightarrow 3H_2(g)$$

Adding: $2Al(s) + 6H^+(aq) \rightarrow 2Al^{3+}(aq) + 3H_2(g)$

**b** For a transfer of six electrons:
$$2Al(s) + 8OH^-(aq) \rightarrow 2Al(OH)_4^-(aq) + 6e^-$$
$$6H_2O(l) + 6e^- \rightarrow 3H_2(g) + 6OH^-(aq)$$

Adding: $2Al(s) + 2OH^-(aq) + 6H_2O(l) \rightarrow 2Al(OH)_4^-(aq) + 3H_2(g)$

6. The addition of acid first neutralises excess alkali and then removes hydroxide ions from the aluminate, allowing a precipitate of aluminium hydroxide to appear. The precipitate dissolves in excess acid.

$$Al(OH)_4^-(aq) + H^+(aq) \rightarrow Al(OH)_3(s) + H_2O(l)$$
$$Al(OH)_3(s) + 3H^+(aq) \rightarrow Al^{3+}(aq) + 3H_2O(l)$$

The addition of sodium hydroxide first neutralises excess acid, and then a precipitate of aluminium hydroxide appears. The precipitate dissolves in excess alkali to form aluminate ions.

$$Al^{3+}(aq) + 3OH^-(aq) \rightarrow Al(OH)_3(s)$$
$$Al(OH)_3(s) + OH^-(aq) \rightleftharpoons Al(OH)_4^-(aq)$$

The addition of sodium carbonate first neutralises excess acid, releasing carbon dioxide. Sodium carbonate is alkaline by hydrolysis:

$$CO_3^{2-}(aq) + H_2O(l) \rightleftharpoons HCO_3^-(aq) + OH^-(aq)$$

so that a precipitate of aluminium hydroxide again appears. However, the concentration of hydroxide ions is never sufficient to redissolve the precipitate.

**EXERCISE 4**  **a** Remove the oxide layer by immersing small pieces of aluminium in copper(II) chloride solution. Rinse in water and dissolve in hot dilute sulphuric acid. Reduce the volume of the resulting solution by boiling and allow the solution to cool. Colourless crystals of $Al_2(SO_4)_3 \cdot xH_2O$ separate on cooling. These can be filtered off, washed and dried.

**b i)**  Relative molecular mass of $Al_2(SO_4)_3 \cdot xH_2O$ $= 54 + 288 + 18x$
$= 342 + 18x$

$\therefore$ % of Al in $Al_2(SO_4)_3 \cdot xH_2O = \dfrac{54}{342 + 18x} \times 100 = 8.1$

$\therefore 8.1 \, (342 + 18x) = 5400$
$2770.2 + 145.8x = 5400$
$145.8x = 5400 - 2770.2 = 2629.8$

$\therefore x = \dfrac{2629.8}{145.8} = \mathbf{18}$

**ii)**  Like other solutions containing non-transition-metal ions, an aqueous solution of $Al_2(SO_4)_3 \cdot xH_2O$ is colourless.

**c**  Aluminium does not dissolve readily in dilute acids because the coating of aluminium oxide on its surface protects it. If the coating is removed the aluminium reacts vigorously with dilute acids, as predicted by the strongly negative $E^{\ominus}$ value.

**EXERCISE 5**  **a**  Boiling point of $BF_3$ = 174 K ($-99\ °C$)
Boiling point of $BCl_3$ = 286 K (13 °C)
Boiling point of $AlF_3$ = 1530 K (1257 °C) $\Big\}$ These are sublimation temperatures,
Boiling point of $AlCl_3$ = 451 K (178 °C) $\Big\}$ i.e. at ordinary pressure the transition
is solid $\rightleftharpoons$ gas.

**b**  Aluminium fluoride has the greatest ionic character, which is reflected in its high boiling point. The boiling points of the other halides are more in line with compounds with a high degree of covalent character.
Aluminium fluoride is more ionic than the other compounds because the $Al^{3+}$ ion is less polarising than $B^{3+}$, and because $F^-$ is less polarisable than $Cl^-$.

**c**  In ionic character, $AlF_3 > AlCl_3 > AlBr_3 > AlI_3$
The halide ions increase in size from $F^-$ to $I^-$ and thus become more polarisable. This leads to a much greater degree of covalent character in $AlI_3$ compared to $AlF_3$.

**EXERCISE 6**  **a**

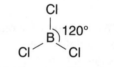

Trigonal planar – all angles 120°

**b**  In view of their considerable covalent character, the chlorides of B and Al would be expected to dissolve in organic solvents.

**c i)**  $M_r$ for aluminium chloride at 400 °C is twice that at 800 °C because at 400 °C the molecules in the vapour exist as dimers, $Al_2Cl_6$, whereas at 800 °C they are completely dissociated into monomers, $AlCl_3$.

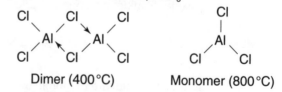

Dimer (400 °C)          Monomer (800 °C)

**ii)**  At 600 °C aluminium chloride vapour consists of the dimer and monomer in dynamic equilibrium with each other.

$$Al_2Cl_6 \rightleftharpoons 2AlCl_3$$

Thus, $M_r$ has a value between that of the dimer and monomer.

**d**  The small size of the boron atom does not permit four chlorine atoms to surround it, i.e. steric hindrance prevents the formation of $B_2Cl_6$ by dative bonding, as in $Al_2Cl_6$. $M_r$ for boron chloride vapour is always the value for the monomer, $BCl_3$, i.e. 117.

**EXERCISE 7**   **a** $BCl_3(l) + 3H_2O(l) \rightleftharpoons B(OH)_3(aq) + 3HCl(aq)$

                          boric acid     hydrochloric acid

        $AlCl_3(s) + 6H_2O(l) \rightleftharpoons [Al(H_2O)_6]^{3+}(aq)$             $+ 3Cl^-(aq)$

                      hexaaquaaluminium(III) ion     chloride ion

**b** The resulting solutions in both cases are acidic.

**c** $BCl_3$ is an 'electron-deficient' compound (only six electrons found around B) and can thus form dative covalent bonds with the oxygen atoms in water. This initiates the hydrolysis reaction. On the other hand, $CCl_4$ is not electron deficient and cannot form similar dative bonds. (Note that hydrolysis of other Group IV chlorides can occur by the use of d orbitals which are not available for carbon – see ILPAC 11, Group IV Elements.)

**d**

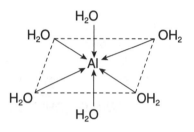

This is an octahedral ion.

**e** No d orbitals are available for boron to form dative bonds with ligands in the way aluminium does. Even if there were, $B^{3+}$ is too small to accommodate six ligands around it.

**f** An alternative molecular formula is $H_3BO_3$.

**EXERCISE 8**   **a** $[Al(H_2O)_6]^{3+}(aq) + H_2O(l) \rightleftharpoons [Al(H_2O)_5(OH)]^{2+}(aq) + H_3O^+(aq)$

        $[Al(H_2O)_5(OH)]^{2+}(aq) + H_2O(l) \rightleftharpoons [Al(H_2O)_4(OH)_2]^+(aq) + H_3O^+(aq)$

**b** For $B(OH)_3$, $K_a(298\ K) = 5.8 \times 10^{-10}\ mol\ dm^{-3}$

    For $[Al(H_2O)_6]^{3+}$, $K_a(298\ K) = 1.0 \times 10^{-5}\ mol\ dm^{-3}$

    $[Al(H_2O)_6]^{3+}$ is a stronger acid than $B(OH)_3$.

**c** In $B(OH)_3$, boron is electron-deficient and can therefore form dative covalent bonds with the oxygen atoms in water.

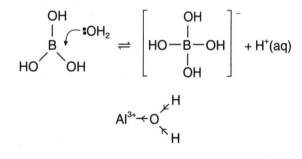

**d**

The O — H bond in the hexaaquaaluminium ion is polarised and weakened by the small and highly charged aluminium ion at the centre of the complex strongly attracting the electrons of the Al — O bonds towards itself. This, in turn, causes the electrons of the O — H bonds to move towards the oxygen atoms and facilitates the loss of protons.

**e** The movement of electrons mentioned in (d) results in a less highly charged aluminium ion in $[Al(H_2O)_5(OH)]^{2+}$ once the first proton has been released. This reduced charge reduces the polarisation of the Al — O bonds. Thus, the O — H bonds in $[Al(H_2O)_5(OH)]^{2+}$ are stronger than in $[Al(H_2O)_6]^{3+}$ and therefore less likely to release protons.

**EXERCISE 9**   **a** In the hydrolysis of $[Al(H_2O)_6]^{3+}$ the following equilibria exist:

$$[Al(H_2O)_6]^{3+}(aq) + H_2O(l) \rightleftharpoons [Al(H_2O)_5(OH)]^{2+}(aq) + H_3O^+(aq)$$
$$[Al(H_2O)_5(OH)]^{2+}(aq) + H_2O(l) \rightleftharpoons [Al(H_2O)_4(OH)_2]^+(aq) + H_3O^+(aq)$$

In the presence of stronger bases than water, the above equilibria are displaced over to the right by the removal of the hydronium ion. Furthermore, a base such as sodium carbonate solution is strong enough to remove one more proton from $[Al(H_2O)_4(OH)_2]^+$ forming an insoluble white precipitate of hydrated aluminium hydroxide.

$$2[Al(H_2O)_4(OH)_2]^+(aq) + CO_3^{2-}(aq) \rightleftharpoons 2[Al(H_2O)_3(OH)_3](s) + CO_2(g) + H_2O(l)$$

**b** In the presence of an excess of strong base, such as aqueous sodium hydroxide, a proton can be removed from each of the co-ordinated water molecules in $[Al(H_2O)_6]^{3+}$. Thus, if the sodium hydroxide solution is not in excess, a white precipitate of aluminium hydroxide is formed.

$$[Al(H_2O)_4(OH)_2]^+(aq) + OH^-(aq) \rightleftharpoons H_2O(l) + [Al(H_2O)_3(OH)_3](s)$$
hydrated aluminium hydroxide

But in the presence of excess hydroxide ions, further protons are removed to produce the soluble aluminate(III) ions:

$$[Al(H_2O)_3(OH)_3](s) + OH^-(aq) \rightleftharpoons [Al(H_2O)_2(OH)_4]^-(aq)^* + H_2O(l)$$
$$[Al(H_2O)_2(OH)_4]^-(aq) + OH^-(aq) \rightleftharpoons [Al(H_2O)(OH)_5]^{2-}(aq) + H_2O(l)$$
$$[Al(H_2O)(OH)_5]^{2-}(aq) + OH^-(aq) \rightleftharpoons [Al(OH)_6]^{3-}(aq) + H_2O(l)$$

**EXERCISE 10**
**a**
Figure 3

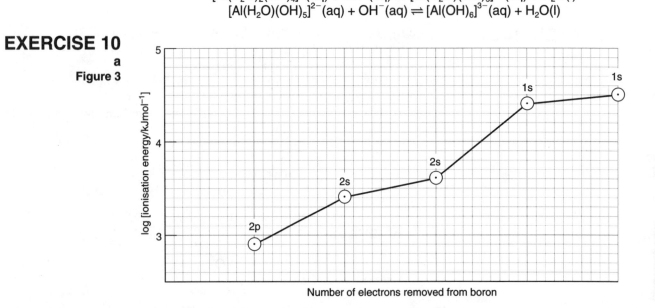

Number of electrons removed from boron

The graph (and the table) show that there are three electrons in the outer shell and, therefore, the most likely formula of boron oxide is $B_2O_3$.

*This form predominates and is usually written as $Al(OH)_4^-$. In some books you may find it written in a totally anhydrous form as $AlO_2^-$ [i.e. $Al(OH)_4^- - 2H_2O$].